To the Reader

If you are considering a study of this book you probably fit into one of the following categories:

- You are not enrolled in a physics course, but you wish to learn, or review, some physics—perhaps to enable you to pass a test which requires a knowledge of physics.
- You are taking an introductory physics course and wish to have a study guide to accompany your textbook.
- You are enrolled in a course which uses a nonmathematical, conceptual book as the primary text, but your instructor wishes to include more problem-solving in the course.
- You are enrolled in a course which uses this book as its primary text.

In any case, this Guide *Basic Physics* can help you. It is a complete, self-contained physics book with a programmed format. The chapters are divided into short steps called frames. Each frame presents some new material, and then asks you questions to test your comprehension. By faithfully answering the questions (preferably by actually writing in the answers) before you check the answers I give, you will be able to check your progress continually. In fact, I suggest that you cover the furnished answers with a card until you have completed your own. If your answer does not agree with the one following the dashed line, be sure you understand why before proceeding to the next frame. To check yourself further, a Self-Test—with answers—is included at the end of each chapter.

Since physics builds from one principle to another, many chapters require an understanding of previous chapters. For this reason the prerequisites of each chapter are listed both in the table of contents and on the first page of the chapter. Some frames within chapters, however, may be skipped without disrupting the train of thought. Such frames, which are often mathematical treatments of the subject at hand, are labeled as "optional" and indicated with a vertical line in the margin. Some optional frames have a prerequisite of a prior optional frame; such prerequisites are listed at the beginning of the frame.

I suggest that you complete the entire Self-Test at the end of a chapter before checking answers. In this way your test will be more similar to classroom testing. Each Self-Test answer includes references to the frame(s) to which you should return for help on missed items.

Have fun!

BASIC PHYSICS

BASIC PHYSICS

KARL F. KUHN, Ph.D.
Department of Physics
Eastern Kentucky University

John Wiley & Sons, Inc., Publishers
New York • Chichester • Brisbane • Toronto • Singapore

Editors: Judy Wilson
Production Manager: Ken Burke
Editorial Supervisor: Winn Kalmon
Artist: John King

Library of Congress Cataloging in Publication Data:

Kuhn, Karl F.
    Basic physics.

    (Wiley self-teaching guides)
    Includes index.
    1. Physics. I. Title
QC23.K74   530   78-23384
ISBN 0-471-03011-2

Printed in the United States of America

79 80 10 9 8 7 6 5 4

DEDICATION

To the Designer of this beautiful universe,
in thanksgiving for making it comprehensible*

---

*"The most incomprehensible thing about the universe is that it is comprehensible."
—Albert Einstein

ACKNOWLEDGMENTS

First I want to thank Chuck Lasher, the John Wiley & Sons representative who encouraged me to write the book. He's a truly good man. Thanks are certainly due my reviewers, particularly Jesse David Wall, Paul Hewitt, and C. Sherman Frye, for their very helpful comments and suggestions. Ruth Ashley did so much quality reviewing—make that "rewriting"—that she should be listed as co-author. Judy Wilson, the Super STG Editor, had the job of coordinating and controlling the whole endeavor. She and Winn Kalmon (our California connection) put it all together. Only those who saw my original sketches can truly appreciate John King's artwork. Finally, love and thanks to my family: my beloved wife Sharon and our "kids," Karyn, Kim, Karl Jr. (Deuce), Keith, and Kevin. They put up with a lot.

Contents

PART ONE

The Physics of Force and Motion

The science which we now call physics started, in its modern form, with Galileo Galilei in the sixteenth century. The primary subject of his concern was the description of motion. Today this forms the starting point for that area of physics called mechanics, and it is with this subject that most introductory physics books begin.

Here in Part I we will start by looking at the interaction of force and motion. This will lead us quickly to a consideration of energy, which is one of the most valuable concepts in science and which will be with us throughout the book. Then after studying momentum, we end the section by looking at gravity and weightlessness.

CHAPTER ONE

Force and Motion

Physics deals with quantities that can be measured. Thus you won't find concepts such as honesty, love, and courage as primary topics of discussion in a physics book. One notable thing about measurable quantities is that anything you can think of that is measurable can be specified in terms of only four basic dimensions: mass, length, time, and electric charge. In this chapter we will begin to consider the first three of these.

When you have completed your study of this chapter, you will be able to:

- define speed;

- calculate speed, distance, or time, given the other two values;

- differentiate between speed and velocity;

- distinguish between scalar and vector quantities, giving an example of each;

- calculate acceleration, given initial and final velocities and the elapsed time;

- state the value of the acceleration of gravity in both the metric and British systems of units*;

- given an initial velocity, calculate the distance an object will fall in a given time;

- state and give an example of the application of each of Newton's laws of motion;

- distinguish between mass and weight;

- given two of the three quantities—force, mass, and acceleration—calculate the third;

- calculate the weight of an object of known mass;

- use the graphical method to add vectors;

- use Newton's Second Law to explain why objects of different mass fall with the same acceleration;

- specify the cause of terminal speed;

---

*The British system of units is the one you are probably most familiar with. It includes such units as inches, feet, and pounds.

- give an example of an object moving with constant speed while at the same time accelerating;

- relate Newton's second law to circular motion;

- identify the force-pair of Newton's third law for cases of both accelerated and nonaccelerated motion;

- identify and classify (as to acceleration and velocity) each of the two components of projectile motion.

SPEED

1.   Speed is a combination of distance and time; it is defined as the distance traveled divided by the time of travel. Thus, if you travel 12 miles in 3 hours, the average speed is 4 miles per hour (mi/hr or mph).

(a)   If an elephant runs for half an hour at a speed of 6 miles per hour, what distance does it cover? _____

(b)   What is the average speed of a vehicle that covers 250 miles in two-and-a-half hours?

_____

(c)   Using the symbols s, d, and t, write a formula for speed. _____

— — — — — — — — — — — — — — —

(a) 3 miles; (b) 100 mph; (c) $s = d/t$

VELOCITY—A VECTOR QUANTITY

2.   Often speed does not tell us all we want to know about a motion. If you were told that while you slept your sheepdog left you at a speed of 5 mph, and that he traveled in a straight line, you would not have enough information to go looking for him. You would also need to know which way he went. The term *velocity* specifies both speed and direction. Identify the two descriptions below as speed or velocity.

(a)   20 meters/second _____

(b)   20 meters/second northeast _____

(c)   How does velocity differ from speed? _____

— — — — — — — — — — — — — — —

(a) speed; (b) velocity; (c) velocity includes a direction

3.   The defining equation for velocity is:

$$\vec{v} = \frac{\vec{d}}{t}$$

Here $\vec{v}$ is the velocity and $\vec{d}$ is the straight-line distance from the object's starting position to its position after time t has elapsed (d is called the object's displacement). The arrows are placed above some of the symbols to indicate that the quantities are *vectors*—that is, they have a direction as well as a magnitude (size).

(a)   Which quantities in the equation above are vectors? _____

(b)   Which are not vectors? _____

– – – – – – – – – – – – – – – –

(a) velocity and displacement ($\vec{v}$ and $\vec{d}$); (b) time (t)

4.   We cannot assign a special direction to time, so it is not a vector. All quantities that are not vectors are called *scalars.* Identify whether each of the following represents a vector or a scalar quantity.

(a) velocity _____   (b) speed _____   (c) time _____

(d) displacement _____

– – – – – – – – – – – – – – – –

(a) vector; (b) scalar; (c) scalar; (d) vector

5.   Suppose a jogger leaves his house and jogs north 300 feet to the streetlamp on the corner. He turns west, jogs another 400 ft, and gets to the oak tree 2 minutes after leaving home. Use the equation $\vec{v} = \vec{d}/t$ to calculate the jogger's average velocity for the two minutes.

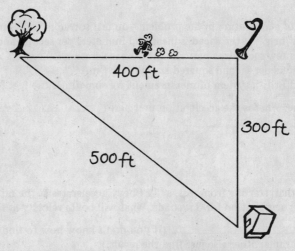

400 ft

300 ft

500 ft

(a)   Which of the three distances given in the figure is the displacement d? _____

(b)   What is the jogger's average velocity? _____

(c)   What is the total distance (not displacement) traveled by the jogger? _____

(d)   What is the average speed (not velocity) during the two minutes? _____

– – – – – – – – – – – – – – – –

(a) 500 feet; (b) 250 ft/min in a direction between North and West; (c) 700 feet; (d) 350 ft/min

ACCELERATION

6.   As a car gains speed, we say that it accelerates. The change in velocity divided by the time it takes to make the change is called the *acceleration*. In equation form:

$$\vec{a} = \frac{\vec{v}_1 - \vec{v}_0}{t}$$

Here $\vec{v}_0$ represents the velocity at the start and $\vec{v}_1$ is the velocity after time t has passed.

Notice that although in everyday language the word "acceleration" is used only to refer to a gain in speed, if $\vec{v}_0$ is larger than $\vec{v}_1$, the acceleration is negative and is actually a deceleration. In physics we use the word "acceleration" to include this negative case.

(a)   Is acceleration a vector or a scalar quantity? _____

(b)   What part of the equation represents the change in velocity? _____

(c)   If a car takes 5 seconds to increase its velocity from 30 miles/hour to 50 miles/hour

(both in the same direction) what is its acceleration? _____ miles per hour per second.

— — — — — — — — — — — — — —

(a) vector; (b) $\vec{v}_1 - \vec{v}_0$; (c) 4

(Note: 30 mi/hr and 50 mi/hr are often called "velocities," even though there is no mention of direction, when the direction of both is the same. Strictly speaking, direction should always be included in stating a velocity.)

7.   The unit of acceleration in the problem you just solved is "miles per hour per second," or "mi/hr/s." If the speed is measured in meters per second and its change is measured over a few seconds, a natural unit for acceleration is "m/s/s," which is sometimes called meters per second squared and written "m/s^2."

In our old British system of measurement we sometimes use feet for length with seconds for time. What is the acceleration unit here? _____

— — — — — — — — — — — — — —

ft/s/s or ft/s^2

8.   Suppose that, starting from rest, a '71 Chevy accelerates at the rate of 5 mi/hr/s and it continues this acceleration for 8 seconds. What will be its velocity at the end of the

eight seconds? _____ (If you don't know how to find the answer, read the rest of this frame, otherwise just find the result.)

First write the acceleration equation:

$$\vec{a} = \frac{\vec{v}_1 - \vec{v}_0}{t}$$

In this case, $\vec{v}_0$ is zero since the car started from rest. Now solve the equation for $\vec{v}_1$, the unknown velocity:

$$\vec{v}_1 = \vec{a}t$$

Substitute in the acceleration and time, and solve for $\bar{v}_1$.

_____

– – – – – – – – – – – – – – – – –

40 mi/hr

0s
0 ft/s

1s
32 ft/s

2s
64 ft/s

3s
96 ft/s

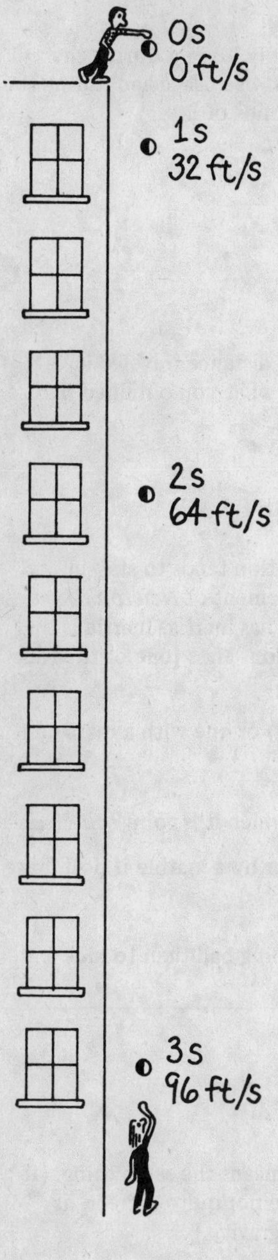

Figure 1.1

ACCELERATION OF GRAVITY

9.   The acceleration of an object falling freely near the surface of the earth is 9.8 m/s^2 or 32 ft/s^2. An object dropped from a very high platform achieves a speed of 9.8 m/s by the end of the first second and 19.6 m/s by the end of the next second.

Suppose an object starts from rest and falls freely. After one second its velocity is 32 ft/s.

(a)   What is its velocity in ft/s after two seconds?

_____

(b)   What is its velocity after three seconds?

_____

– – – – – – – – – – – – – – – – –

(a) 64 ft/s; (b) 96 ft/s

10.   Figure 1.1 shows the speeds at various times of the ball which the boy has dropped from the building. Now suppose that a ball is thrown upward at the starting velocity of 96 ft/s. On the way up its acceleration (or deceleration if you wish) would be 32 ft/s^2, just as it was on the way down, except now the acceleration is downwardly directed while the velocity is upward. Because the acceleration is the same whether the object is moving up or down, Figure 1.1 could just as well represent the girl throwing the ball upward to the boy. In this case, what would be the speed of the

ball when it reached the top? _____

– – – – – – – – – – – – – – – – –

0 ft/s

ACCELERATION EQUATIONS

11.   The equation relating acceleration, the distance traveled, and the time of fall is:

$$\bar{d} = \bar{v}_0 t + \tfrac{1}{2} \bar{a} t^2$$

where $\bar{v}_0$ is the original velocity and t represents time elapsed. Thus, in Figure 1.1, when the ball has fallen

1 second (and its speed is 32 ft/s), it has fallen 16 feet ($\vec{d} = \frac{1}{2} 32 \cdot 1^2$, since $\vec{v}_0$ was 0).
After 3 seconds of all how far has it fallen? _____

----

144 ft ($\vec{d} = \frac{1}{2} \vec{a} t^2 = \frac{1}{2} (32)(3)^2 = \frac{1}{2} (288)$)

12.   Although its speed after 1 second is 32 ft/s, the ball fell only 16 feet during the second. This is because it started from rest—at zero speed—so its *average* speed during the first second is 16 ft/s, or half of 32. Consider the first three seconds of fall.

Initial speed = zero

(a)   Final speed: $\vec{v}_1 = \vec{a} t =$ _____

(b)   Average speed = _____

(a) 96 ft/s; (b) 48 ft/s
(Since it fell for three seconds at an average speed of 48 ft/s, its distance traveled is "48 ft/s times 3 seconds," which equals 144 ft. This corresponds to what you obtained by use of the equation $\vec{d} = \vec{v}_0 t + \frac{1}{2} \vec{a} t^2$.)

MASS AND INERTIA: NEWTON'S FIRST LAW

13.   An object at rest tends to stay at rest and an object in motion tends to stay in motion in a straight line at a constant speed. The above is a statement of *Newton's first law*, and it expresses what we mean when we say that an object has inertia. Inertia is that property of matter which causes matter to resist change in motion. Mass (one of the four basic measurements) is a measure of inertia.

(a)   Which has more inertia, an object of mass 6 kilograms (kg) or one with a mass of

3 kg? _____

(b)   Which has more of a tendency to remain at the speed at which it is going? _____

(c)   Which is more likely to hurt, being hit by a bowling ball or by a marble if both have

the same speed? _____

(d)   Why does it hurt a barefooted person more to kick a bowling ball than to kick a

marble? _____

(a)   6 kg
(b)   6 kg
(c)   bowling ball
(d)   The bowling ball has more mass—or more inertia, which means the same thing. (If you said it hurts more because it has more weight, you are not quite correct, as should become clear after more discussion of weight and gravity.)

NEWTON'S SECOND LAW

14. If one object has more inertia than another, it requires more force to give it the same acceleration. Suppose that you want an object to accelerate at some specific rate, such as 5 m/s^2. The amount of force you must apply depends upon the mass of the object. In fact, if you have an object with twice the mass of another one, you must use double the force to get the same acceleration for the first object. And for three times the mass, you need three times the force. You can see that, for a given mass, force and acceleration are proportional.

The relationship between force, mass, and acceleration is summed up in a single equation, Newton's second law:

$$\vec{F} = m\vec{a}$$

where $\vec{F}$ is force, m is mass, and $\vec{a}$ is acceleration.

(a)   Is force a vector or a scalar? _____

(b)   Which would require a greater force—accelerating a 10 gram mass at 5 ft/s^2 or a 20 gram mass at 2 ft/s^2? _____

– – – – – – – – – – – – – – – –

(a) vector; (b) 10 gram mass at 5 ft/s^2

15. The vector nature of the force equation tells us that the directions of the force and the resulting acceleration are exactly the same. We saw this when we considered the boy dropping the ball from the roof. It was pointed out at that time that the acceleration was downward whether the ball was dropped by the boy or thrown upward by the girl. This was because the force causing the acceleration was the force of gravity, which is always downward.

(a)   Thus if the accelerating force is downward, the acceleration is _____.

(b)   If the accelerating force is to the southeast, what is the direction of the resulting acceleration? _____

– – – – – – – – – – – – – – – –

(a) downward; (b) southeast

Units Used in Newton's Second Law

16. If mass is expressed in kilograms and acceleration in m/s^2, the correct unit of force is the *newton* (N). A newton is about $\frac{1}{4}$ pound, the weight of a stick of margarine. Use the second law equation to solve these problems.

(a)   What is the acceleration produced by a force of 12 newtons exerted on an object of 3 kilograms mass? _____

(b)   To produce an acceleration of 4 m/s^2 on a bowling ball with a mass of 6 kg, what force would be required? _____

(c)   Suppose one wishes to accelerate a 1000 kg Chevy from 10 m/s (which is about 20 mi/hr) to 30 m/s (about 60 mi/hr) in 8 seconds. What force would be required?

_____

— — — — — — — — — — — — — — — —

(a) 4 m/s^2; (b) 24 N; (c) 2500 N
If you are unable to work problem (c), continue below. If you got the answer, go ahead to frame 18.

17.   To solve problem (c) above, work through these steps:

(a)   Calculate the acceleration involved in changing the speed from 10 m/s to 30 m/s in

8 seconds. _____

(b)   Use the formula $F = ma$ to find the force required to accelerate a 1000 kg object

at that rate. _____

— — — — — — — — — — — — — — — —

(a) 2.5 m/s^2; (b) 2500 N

18.   Suppose you observe a 1 kg object accelerating at the rate of 9.8 m/s^2. You know that a force of 9.8 N must be accelerating the object. This acceleration of 9.8 m/s^2 is the acceleration which occurs when an object is released near the surface of the earth—the acceleration of gravity mentioned in an earlier section. Thus the force of gravity on a 1 kg object is 9.8 N. We say that a 1 kg object *weighs* 9.8 newtons.
    The foregoing allows us to state a general method for finding the weight of an object whose mass is known. In your imagination drop an object of mass 10 kg.

(a)   How much will its acceleration be? _____

(b)   What is the resulting force that causes the acceleration? _____

(c)   What is the weight of the object? _____

— — — — — — — — — — — — — — — —

(a) 9.8 m/s^2; (b) 98 N; (c) 98 N

19.   We have seen a definite distinction between weight and mass. Weight is a force which is due to the pull of gravity of the earth. Mass is a measure of the inertia possessed by an object. Mass is the more fundamental quantity because the amount of inertia of an object does not depend upon its location relative to the earth, but the weight of an object does—an object weighs less as it gets farther and farther from the earth.

(a)   Which system, metric or British, commonly uses mass rather than weight?

_____

(b)   Which is more stable—the mass or the weight of an object? _____

— — — — — — — — — — — — — — — —

(a) metric; (b) mass

Adding Forces

20.   In the discussion of forces above, we were referring to unbalanced forces on the object—those forces responsible for the acceleration. Suppose a 2 pound banana is resting on a plate. The banana weighs 2 pounds, which means that gravity is pulling downward on it with a force of 2 pounds. The plate, however, is pushing upward on it equally hard. Thus the banana's weight is balanced by another force. The two forces on the banana are equal in magnitude (size) but opposite in direction, so they *add* to zero.

The force referred to in Newton's second law is the resultant, or *net*, force on an object. Suppose two kids want a toy truck. One kid pulls one way with a force of 5 pounds and the other pulls in the opposite direction with a force of 4 pounds.

(a)   What is the net force on the toy? _____

(b)   What force causes the toy to accelerate? _____

— — — — — — — — — — — — — — — —

(a) 1 pound; (b) the *net* force of 1 pound

21.   When all the forces on an object total zero, the object is said to be *in equilibrium*. In everyday language we often use the word equilibrium to refer only to objects at rest, but an object in motion can also be in equilibrium. As a car travels at a constant speed on a level, straight road, the engine causes a force to be exerted forward on the car. This force, if the car is not accelerating, is exactly balanced by the forces of friction on the car. All the forces on the car total zero. This may seem odd at first, but if you remember Newton's first law, you'll recall what happens to a moving object when no unbalanced

force acts on it. Describe the state of motion of such an object. _____

_____

— — — — — — — — — — — — — — — —

It continues at a constant speed in a straight line. (It does *not* stop, for this would imply acceleration and acceleration requires an unbalanced force.)

VECTOR ADDITION

(1)

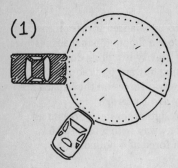

22.   If two or more forces are exerted on an object, we must take into account their directions in order to add them. This is most easily accomplished by using scale drawings (the "graphical" method). Suppose two cars push on the huge pie of Figure 1.2, as shown in part 1. Part 2 uses arrows to represent the forces they exert. Since the Volkswagen exerts 160 pounds of force and the black Ford 80, the arrow representing the VW's force is twice as long as the Ford's (4 centimeters and 2 cm respectively). In part 3, the arrows are hooked end-to-end, keeping each the same length and pointed in the same direction as the forces exerted.

Now *you* draw a line from the tail of the first arrow (the long one) to the head of the second one. This vector will represent the net force, so place an arrowhead at its upper end. Go ahead and draw it now, then measure it. What is its length? _____

(2)   80 lb (2 cm)

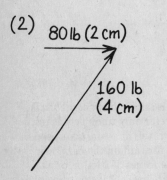

160 lb (4 cm)

(3)

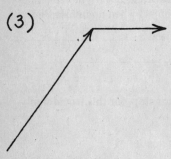

Figure 1.2

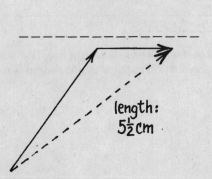

length: 5½ cm

23.   Since we used 1 cm to represent 40 pounds, your resultant vector represents a force of 220 lb ($5\frac{1}{2}$ times 40) pointing in a direction between the two cars' directions of force. Note that it is, as should be expected, a little closer to the direction the VW is pushing. (This is because the VW is pushing much harder and thus affects the resultant direction more than does the Ford.)

(a)   Using the same scale (1 cm representing 40 lb) use the space below to draw the vector diagram if the VW pushes with a force of 140 lb and the Ford with 200 lb.

(b)   What is the resultant force in this case? _____

(c)   Since direction is considered in adding forces, what type addition is it? _____

– – – – – – – – – – – – – – – –

(a)

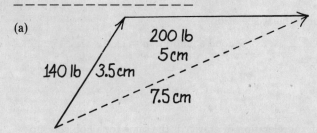

200 lb
5 cm

140 lb / 3.5 cm

7.5 cm

(b)   300 lb (If your resultant vector measures between 7.0 and 8.0 cm, that is OK. You should get between 285 and 320 lb.)

(c)   vector

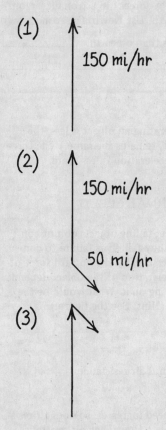

(1)   150 mi/hr

(2)   150 mi/hr

50 mi/hr

(3)

Figure 1.3

24.   Vector addition also is used with velocities. Consider an airplane flying in still air at a velocity of 150 miles per hour northward. In Figure 1.3 (part 1) a 3 cm arrow represents this velocity. Now a wind starts blowing toward the southeast at a speed of 50 mi/hr—quite a wind! Figure 1.3 (part 2) shows this 50 mi/hr wind as a 1-cm arrow. Just as was done in the addition of forces, one arrow is moved so that its tail falls on the head of the other, shown in part 3.

(a)   Draw the resultant vector in part 3.

(b)   What is its length? _____

(c)   What net force does this represent?

_____

– – – – – – – – – – – – – – –

(a)   It is drawn from the tail of the first to the head of the second.

(b)   2.4 cm

(c)   120 mi/hr (because each cm represents 50 mi/hr) (Note that the plane is blown off course as well as being slowed down.)

GRAVITY AGAIN

25.   If an object (call it object A) has twice as much mass as another object (B), it also weighs twice as much; the force of gravity on it is twice as much. Consider each of these objects falling under the influence of only one force, gravity, and let's apply Newton's second law.

$$F = ma$$

Now the heavier object (B) has twice the mass, or $2m$. But it also has twice the weight, or $2F$. Substituting these symbols for m and F, we get:

$$2F = (2m)a$$

Solve each of these equations for a, the acceleration.

First equation: $a =$ _____ ;   Second equation: $a =$ _____

— — — — — — — — — — — — — — —

$a = F/m$ in both cases

26.   This result explains why the acceleration of gravity is the same for all objects, no matter what their weight. We give the acceleration of gravity a special symbol, g.

$$g = 32 \text{ ft/s}^2$$
$$= 9.8 \text{ m/s}^2$$

(You'll see in Chapter 4 that this is the origin of the term "g-forces" in astronaut jargon.)
    Objects on the moon weight less than they do on earth. Use Newton's second law to

explain why the acceleration of gravity on the moon is less than on earth. _____

_____

_____

— — — — — — — — — — — — — — —

From the equation $F = ma$, we see that if the force accelerating an object is less, the acceleration of the object will be less. (The mass, of course, remains the same.) The force of gravity on the moon is less than that on earth, so the acceleration is less there.

Air
resistance

Weight

TERMINAL SPEED

27.   In actual practice, falling objects do not continue to accelerate as they fall. This failure to continue gaining speed is caused by the upward force of air resistance. The acceleration of the object depends upon the net force acting on it. What would be the net force in this case? (Hint: See the figure.)

_____

— — — — — — — — — — — — — —

the object's weight minus air resistance

28.   As downward speed increases, so does air resistance. Thus there comes a time when air resistance equals the weight of the object. At this speed the object no longer accelerates, but instead it continues at the same speed—the terminal speed. (This follows from $F = ma$, since if the net force is zero, the acceleration must also be zero.)

Consider a falling rock at a time when it is traveling just less than terminal speed. Will its acceleration be less than, equal to, or more than 9.8 m/s^2 at this time? _____

— — — — — — — — — — — — —

Less. At terminal speed the air resistance is equal to the rock's weight. Just below this speed there is considerable air resistance, although less than the rock's weight. The net force on the rock is its weight minus air resistance. Since this must be less than its weight, the rock's acceleration must be less than the acceleration of gravity.

CIRCULAR MOTION

29. When an object moves in a circle, there is an unbalanced force on it which is toward the center of the circle. (Consider a rock whirled on a string—the string provides the force toward the center.) This force is called the *centripetal* force.

(a) According to Newton's second law, what happens if an object has an unbalanced

force on it? _____

(b) Since you know it is possible for an object to move around in a circle at a constant

speed, how does the acceleration exist in this case? _____

— — — — — — — — — — — — —

(a) it accelerates (or changes velocity); (b) as a change in *direction*
Remember that acceleration is defined as the change in velocity divided by the change in time, and that velocity includes direction.

30. Suppose the string in the figure below breaks as you whirl the rock.

(X)   (Y)

(a) Will the rock fly off as shown in X or Y? _____

(b) Which of Newton's laws explains this? _____

— — — — — — — — — — — — —

(a) Y; (b) the first (The rock tends to continue in the same direction.)

NEWTON'S THIRD LAW (THE LAST!)

31.   Remember the 2 pound banana on the plate? The earth exerts a gravitational force of 2 pounds downward on the banana, and we said that the plate balances this with a force of 2 pounds upward on the banana. If the plate pushes on the banana, the banana *necessarily* exerts an equal and oppositely-directed force on the plate. These two forces form the force-pair of Newton's third law:

> For every force that object A exerts on object B, there is an equal and opposite force exerted by B on A.

In addition to the two forces on the banana—the downward force, *a*, of gravity on the banana (which is its weight) and the upward force, *b*, of the plate on the banana—there are three forces exerted on the plate. These are the force *c* of gravity on the plate (which is the weight of the plate), the downward force *d* of the banana on the plate, and the force *e* exerted upward on the plate by the table or floor.

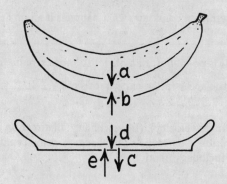

(a)   Which two forces form the action-reaction pair of Newton's third law? _____

(b)   State each of Newton's three laws. _____

_____

_____

_____

— — — — — — — — — — — — — — — —

(a)   *b* and *d*
(b)   1.   An object tends to continue at the same velocity.
      2. $F = ma$
      3.   For every force object A exerts on B, B exerts an equal and opposite force on A.

32.   Although forces *a* and *b* in the preceding frame are equal and in opposite direction, they are not the forces referred to in Newton's third law. Newton's law involves (1) the force one object exerts on a second and (2) the force the second exerts on the first. The tree limb in the figure at the top of page 17 exerts a 12 ounce upward force on the apple (which weighs 12 oz.). What is the reaction force to the force exerted by the limb?

_____

_____

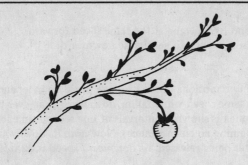

_____

The apple exerts a 12 ounce downward force on the limb.

Newton's Third Law During Acceleration

33.   The third law states that for _every_ force exerted by A on B, there is an equal and opposite force by B on A. If this is true, there must be a force equal and opposite to the apple's weight. To see what this opposite force is we must ask what exerts the gravitational force on the apple. The earth does. Thus the reaction force is the force of gravity exerted by the apple on the earth. Think of gravity as being a stretched spring between the apple and the earth. There is a force downward on the apple and upward on the earth.

But what if the apple falls? While it is falling, Newton's third law still applies. The only force then acting on the apple is the force of gravity. Since it is not balanced by another force, the apple accelerates—according to Newton's second law. The apple likewise continues to exert a force upward on the earth. Why, then, don't we see the earth rise to meet the falling apple? (Hint: Apply Newton's second law to the earth.)

_____

_____

_____

_____

The force here is equal to the weight of the apple—a few newtons. However, the mass of the earth is so great that the resulting acceleration is insignificant. (But we have no reason to think that the law no longer applies, just because its effect is too small to measure. By the very nature of forces, Newton's laws hold in all cases.)

Projectile Motion

34.   Imagine an airplane dropping a hockey puck. If there is no air resistance, the puck is moving forward at the speed of the plane when it is dropped.

(a)   What does Newton's first law tell us about the hockey puck's motion? _____

_____

(b)   What effect will gravity have on the hockey puck? _____

_____

_____

(a)   the puck will tend to continue at the same speed forward

(b)   gravity will cause the puck to accelerate downward

35.   Amazingly, the two motions of the dropped puck are independent: the puck continues forward at the same speed while falling downward at an ever-increasing speed. In
the figure below the puck is shown at intervals of one second. In each second it moves
forward 100 feet (assuming no air resistance). Now note its downward motion. At the end
of the first second after being released, its *downward* speed is 32 ft/s.

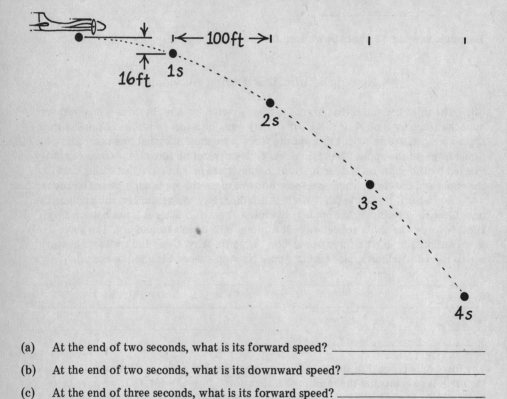

(a)   At the end of two seconds, what is its forward speed? _____

(b)   At the end of two seconds, what is its downward speed? _____

(c)   At the end of three seconds, what is its forward speed? _____

(d)   At the end of three seconds, what is its downward speed? _____

— — — — — — — — — — — — — — — —

(a) 100 ft/s; (b) 64 ft/s; (c) 100 ft/s; (d) 96 ft/s

36.   At the end of three seconds, the puck will have moved 144 ft downward, as well as
300 ft forward. In actual practice, as the puck continues to fall, air resistance will have an
effect.

(a)   What will be its effect on the forward speed? _____

_____

(b)   What will be its eventual effect on the downward speed?_____

_____

_ _ _ _ _ _ _ _ _ _ _ _ _ _ _ _

(a)   it would decrease
(b)   the puck will eventually reach the terminal speed

VECTOR APPLICATION

37.   Let us assume that at the end of the fourth second the forward speed of the puck
has slowed to 80 ft/s and that the puck has reached a downward speed of 90 ft/s (rather
than the 128 ft/s which would occur if air resistance were not present). Using the graphical method of vector addition, determine the actual speed of the puck at this time.

_ _ _ _ _ _ _ _ _ _ _ _ _ _ _ _

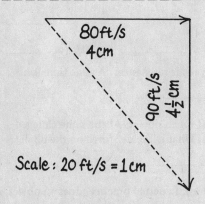

The arrow representing the total is 6.0 cm long, representing 120 ft/s.

SELF-TEST

The questions below will test your understanding of this chapter. Use a separate sheet of
paper for your diagrams or calculations. Compare your answers with the answers provided
following the test.

1.   Define speed. _____

2.   Distinguish between speed and velocity. _____

3.   A train goes 300 miles at an average speed of 45 miles per hour. How long does it take to go the distance? _____

4.   Is acceleration a scalar or a vector quantity? _____

5.   The acceleration of gravity in British units is _____ . The acceleration of gravity in metric units is _____ .

6.   State Newton's first law. _____
_____

7.   State Newton's second law in equation form. _____

8.   One boy pulls on his Radio Flyer (a wagon) with a force of 15 pounds. His (former) friend pulls on it with a force of 18 pounds in the opposite direction. What is the net force on the wagon? _____

9.   Explain why a heavier object does not fall faster when dropped, since the accelerating force (its weight) is greater. _____
_____

10.   My weight is the result of the earth's gravitational force on my body. What is the corresponding reaction force? _____

11.   An object falling from rest reaches a speed of _____ ft/s (or _____ _____ m/s) after falling 5 seconds. (Ignore air resistance.) At this time it has fallen _____ ft (or _____ m).

12.   A force of 3 pounds is exerted toward the east on an object. At the same time a 4 pound force toward the north is exerted on it. What is the net force on the object? Describe the direction of this force. _____

13.   The acceleration of gravity is 32 ft/s^2 downward. In actual practice, does an object continue to accelerate if dropped from a tall building? Explain. _____
_____

14.   What is the reaction force to the weight of this book? _____
_____

15.   Three newtons are exerted northward on a 10 kg object while 4 newtons are exerted southward. What is the acceleration of the object? _____

16.   A 3 kg object weighs how much? _____

Answers to Self-Test

If your answers do not agree with those given below, review the frames indicated in parentheses before you go on to the next chapter.

1. distance traveled divided by the time used to cover that distance   (frame 1)

2. Velocity includes direction as well as speed.   (frame 2)

3. $6\frac{2}{3}$ hours
Solution:

$$s = \frac{d}{t}$$

$$t = \frac{d}{s} = \frac{300 \text{ miles}}{45 \text{ mi/hr}} \quad \text{(frame 1)}$$

4. vector   (frame 6)

5. 32 ft/s^2; 9.8 m/s^2 (frame 9)

6. An object at rest tends to stay at rest and an object in motion tends to stay in motion in a straight line at a constant speed. (Or: An object tends to keep the same velocity.)   (frame 13)

7. $F = ma$ (frame 14)

8. 3 pounds toward the former friend   (frame 20)

9. Its mass is greater by just the same ratio as its weight. Thus, an application of Newton's second law ($F = ma$) results in the same acceleration.   (frame 25)

10. my body's gravitational force on the earth   (frames 31–32)

11. 160 ft/s (49 m/s); 400 ft (122.5 m)   (frames 6, 11, 12)

12. 5 pounds. The direction is east of north, or if you have a protractor you can measure it to be $37°$ east of north. See the figure at the right. (frame 22)

13. No. Because of air resistance, the object reaches a terminal speed.   (frames 27–28)

14. the pull of gravity upward on the earth by the book   (frame 31)

15. 0.1 m/s^2 southward (Solution: The resultant of the forces is 1 newton southward. Using $F = ma$, the acceleration a is 0.1 m/s^2.) (frames 16, 20)

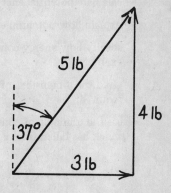

16. 29.4 N (Solution: If dropped, the acceleration would be 9.8 m/s^2. Application of Newton's second law yields the force on the object, which is its weight.) (frames 16, 18)

CHAPTER TWO

Energy and Its Conservation

Prerequisite: Chapter 1

One of the most important concepts in the field of physics is energy. In fact, physics has been defined as the study of matter and energy and their interactions. The word "energy" is part of our everyday vocabulary, but you'll find that in physics the word has a more specific, definite meaning than is normally given.

After completing this chapter, you will be able to:

- define work and power;

- given force, distance, and time, calculate work and power;

- state correct units for work and power in both the metric and British systems;

- define and differentiate between potential energy and kinetic energy;

- calculate potential energy, given weight (or mass) and height;

- explain how potential energy, kinetic energy, and work are related in a system;

- specify how energy conservation applies to machines, giving a quantitative example;

- given the dimensions of a lever, and the load to be lifted, calculate the force required;

- use the principle of conservation of energy to calculate the speed of an object which has fallen from a given height (optional).

WORK

1.   In physics, work is defined as the product of the net force exerted and the distance over which that force is exerted, or $W = Fd$. This corresponds to our everyday meaning of the word in that when you lift a brick 6 feet from the floor you do twice as much work as when you lift it 3 feet. Or suppose you lift 50 bricks to a height of 3 feet. You do 50 times as much work as lifting one brick that high.

But now here comes a conclusion which may seem odd. *After* you lift those 50 bricks up to your waist, you are told by your boss to hold them there for five minutes.

Do you do work in holding them there? To answer this we must go back to our definition. The force you exert is simply the weight of the 50 bricks.

(a)   In holding them waist-high, does that force move the bricks through a distance?

_____

(b)   Using the formula, calculate the total work done in holding the bricks. _____

— — — — — — — — — — — — — — —

(a) no; (b) none
(In holding the bricks, there is motion within the muscles, so work is done by your body, but no external work is done on the bricks. It is this external work with which we are concerned.)

2.   If force is measured in pounds and distance in feet, the unit of work is the ft·lb, or foot·pound. In SI units,* recall (from Chapter 1, frame 16) that force is measured in newtons and distance in meters. The unit of work is the newton·meter, called the *joule* (pronounced "jewel").

(a)   Now suppose that a brick weighs 4 pounds. When you lift those 50 bricks to a

height of 3 feet, how much work do you do? _____

(b)   How much work is done in lifting a 3 newton grapefruit 2 meters? _____

— — — — — — — — — — — — — — —

(a) 600 ft·lb; (b) 6 joules

POWER

3.   Power is defined as work done divided by the time used to do the work, or $P = W/t$. Here again an everyday term is defined in a way which makes it measurable. In a given amount of time a powerful engine can do more work than an engine with less power.

If the two engines do the same amount of work, which can do it in less time? _____

_____

— — — — — — — — — — — — — — —

the more powerful one

4.   Based upon the definition, and using seconds as the unit of time, power in the British system is measured in units of ft·lb/s. If an engine can lift 550 lb to a height of one foot in one second, it is said to have a power of one horsepower. Suppose it lifts 1100 lb to a height of 8 feet in 4 seconds.

(a)   How much work does it do in the 4 seconds? _____

(b)   What is its power in ft·lb/s? _____

_____

*These are the French initials for "International System"—the metric system.

(c)   What is its power in horsepower? _____

– – – – – – – – – – – – – – – –

(a) 8800 ft·lb; (b) 2200 ft·lb/s; (c) 4 horsepower

5.   Corresponding to the above, the SI unit of power is the joule/second. Just as we renamed the newton·meter as the joule, we define the joule/s as a *watt*. An example: a particular machine is capable of a power of 1000 watts. How far can it lift a 50 newton weight (about 11 pounds) in 3 seconds? To answer this, combine work and power formulas to get:

$$P = \frac{Fd}{t}$$

(a)   Use your knowledge of algebra to find the formula for d. $d = $ _____

(b)   Solve the example problem. Your answer? _____

– – – – – – – – – – – – – – – –

(a) $d = \frac{Pt}{F}$; (b) 60 meters

ENERGY

So finally we get to the title of the chapter. The reason for the delay is simple: *Energy is defined as the ability to do work.* There are many types of energy, including electrical energy, heat, and nuclear energy. In this chapter we will be concerned primarily with two types of *mechanical* energy: gravitational potential energy and kinetic energy.

Potential Energy

6.   An object can have energy—the ability to do work—because of its position. A weight which is high above your head can be made to exert a force as it is lowered. Since gravity provides this energy, it is correctly called *gravitational potential energy*, but we usually abbreviate this to "potential energy".

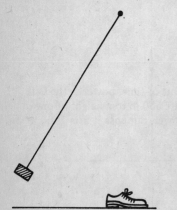

Consider the pendulum in the figure. Release it. When the pendulum hits the shoe, it will push the shoe forward, exerting a force on it over a distance.

(a)   If the force exerted on the shoe is multiplied by the distance the shoe moves, what quantity

is obtained? _____

(b)   The pendulum had the ability to do work on the shoe because when it was at the position shown in the figure it had what?

_____

– – – – – – – – – – – – – –

(a) work; (b) gravitational potential energy

Energy Conservation

7.   A pendulum will always rise to the height from which it fell unless—as when the shoe was there—it does work on something. Now look at the figure to the left, and suppose that the shoe is not there, and ignore friction. Will the pendulum rise to position 1,

2, or 3 at the other end? _____

— — — — — — — — — — — — — — — — —

2

8.   The "something" on which the pendulum does work can be the air. The pendulum exerts a force on air molecules and pushes them aside. This explains why, in actual practice, a pendulum does not rise quite to the height from which it fell. But if the pendulum does not move too fast, it does not exert so much force on the air molecules, and it rises *very nearly* to its original height. Let's consider the potential energy of the pendulum. Since potential energy depends upon height, how would the potential energy at one end of the swing compare to the potential energy at the other end

(assuming no air friction)? _____

— — — — — — — — — — — — — — — —

they are equal (Potential energy is *conserved* in this case.)

Potential Energy Calculations

9.   An object at a given height has an amount of potential energy equal to the work done in raising it to that height. Or, since the force to lift an object is equal to the object's weight:

$$PE = Weight \times Height$$

Consider a 2 pound object that is 6 feet above the floor.

(a)   What is its potential energy with reference to the floor? _____

(b)   How much work can it do when it falls? _____

(c)   Suppose the object falls onto a marshmallow on the floor, and mashes the marshmallow down one inch ($\frac{1}{12}$ foot). What force is exerted by the object on the marshmallow? _____

— — — — — — — — — — — — — — —

(a) 12 ft·lb; (b) 12 ft·lb; (c) 144 lb   Solution:

$$W = Fd$$
$$12 \text{ ft·lb} = F \times \tfrac{1}{12} \text{ ft}$$
$$144 \text{ lb} = F$$

10.   A 5 kg object rests on a table. You lift it up to the ceiling, which is 2 meters above the table. What is its potential energy with reference to the table? (You can calculate this immediately, or follow the steps below.) _____

Remember that the 5 kg is the mass of the object, but that we must find its weight, since its weight is equal to the force we have to exert to lift it. Newton's second law states:

$$F = ma$$

or in this case:

Weight $= mg$ (where g is the acceleration of gravity)
Weight $= 5$ kg (9.8 m/s^2)

Weight $=$ _____ (include units in your answer)

PE $=$ Weight $\times$ Height

PE $=$ _____

– – – – – – – – – – – – – – – –

Weight $= 49$ newtons, since kg (m/s^2) gives newtons
Potential energy $= 98$ joules (or newton $\cdot$ meters)

Kinetic Energy

11.   Consider the pendulum at the left. When it was at position A, it had potential energy, and this energy allowed it to do work on the shoe. In this figure, it has fallen to the level of the shoe and has thus lost its potential energy. So how can it move the shoe? The answer is that the pendulum still has energy, but now the energy is due to its motion rather than to its position. *Kinetic energy* is the energy an object has by virtue of its motion.

(a)   What form of energy is due to position?

_____

(b)   What form of energy is due to motion?

_____

(c)   Which of these energy forms can a pendulum have? _____

– – – – – – – – – – – – – – –

(a) potential; (b) kinetic; (c) both potential and kinetic

12.   The amount of kinetic energy an object has also depends upon the object's mass (as well as its speed). Mass, rather than weight, is the important factor because even if the object were far enough from a planet to be considered weightless, it would still have kinetic energy if it were moving. The formula we use to calculate kinetic energy is:

$$KE = \tfrac{1}{2} mv^2$$

(a)   Would doubling the mass or doubling the velocity of an object have a greater effect

on its kinetic energy? _____

(b)   Suppose the velocity of an object is tripled. How does its kinetic energy change?

_____

_ _ _ _ _ _ _ _ _ _ _ _ _ _ _ _

(a)   velocity, since it is squared in the formula
(b)   it would be 9 times as great ($3^2 = 9$)

Energy Conservation Again
13.   The kinetic energy of the pendulum just before hitting the shoe is equal to the potential energy it lost in falling from the end of its swing. In fact, if we ignore air friction and if the shoe were not there, the potential energy attained on the right end of the swing would be equal to the kinetic energy at the bottom. Now consider the effect of friction, and compare the potential energy at the right of the swing to the potential energy originally at the left.

(a)   If friction is present, how will the potential energy at the right compare to that on

the left? _____

(b)   What will cause this? _____
_ _ _ _ _ _ _ _ _ _ _ _ _ _ _ _

(a) it will be less; (b) work done on the air molecules

14.   The conservation of energy applies during the entire fall of an object. At any time, the potential energy which has been lost equals the kinetic energy which has been gained. Stated another way, the total of the kinetic and potential energy is a constant. We will see later in the book that the law of conservation of energy includes all types of energy. The law can be stated as follows:

> The total energy at the end of
> any event is equal to the total
> energy before the event.

> Or

> Energy can be neither created
> nor destroyed.

(If you are concerned with the energy that is "lost" when there is friction, we will see that when this occurs, heat, a form of energy, is produced. So energy is indeed conserved.)

(a)   At the point where the pendulum has lost 9 joules of potential energy, how much

kinetic energy has it gained? _____

(b)   How much potential energy is needed to develop 15 ft·lb of kinetic energy?

_____

(c)   Suppose the pendulum has 20 joules of PE at the top of its swing. When the pendulum is part way down, its PE is 12 joules. What is its KE at this point?

_____

_ _ _ _ _ _ _ _ _ _ _ _ _ _ _ _

(a) 9 joules; (b) 15 ft·lb; (c) 8 joules

Optional   MORE ENERGY CALCULATIONS

15.   A man drops a 10 kg rock from the top of a 5-meter ladder. What kinetic energy does it have when it hits the ground? Calculate the answer yourself if you can. Otherwise use the steps below. (Hint: First calculate the potential energy at the top.)

$$\text{Potential energy at top} = \text{Weight} \times \text{Height}$$

$$PE = mg \times \text{Height (Remember, weight} = mg, \text{ from Chapter 1, frames}$$
$$25\text{--}26)$$

(a) $= \underline{\quad} \text{ kg} \cdot \underline{\quad} \text{m/s}^2 \cdot \underline{\quad} \text{m}$

(b) $= \underline{\quad}(\text{kg} \cdot \text{m/s}^2) \cdot \text{m}$

(c) $\cdot = \underline{\qquad\qquad} \cdot \text{m}$

(d) $= \underline{\qquad\qquad}$

- - - - - - - - - - - - - - - -

(a) 10 kg · 9.8 m/s² · 5 m; (b) 490; (c) 490 newtons·m; (d) 490 joules
Since we are assuming that no energy was lost to frictional forces, this potential energy at the top is equal to the kinetic energy at the bottom.

16.   What is the speed of the rock in frame 15 just before it hits the ground? _____

- - - - - - - - - - - - - - - -

10 m/s. Solution: $KE = \frac{1}{2} mv^2$

$$490 = \frac{1}{2} \cdot 10 \cdot v^2$$

$$v^2 = 98$$

$$v \cong 10 \text{ m/s} \ (\cong \text{ means "approximately equal to")}$$

17.   If the rock of frame 16 does 85 joules of work against air friction, what is its speed at the bottom? _____

- - - - - - - - - - - - - - -

9 m/s. Solution: $PE_{top} - \text{Work Done} = KE_{bottom}$

$$490 \text{ joules} - 85 \text{ joules} = \frac{1}{2} mv^2$$

$$405 \text{ joules} = \frac{1}{2} \cdot 10 \text{ kg} \cdot v^2$$

$$v^2 = 81$$

$$v = 9 \text{ m/s}$$

18.   A 1000 kg car traveling at 30 m/s (which is about 60 miles/hour) stops in a distance of 50 meters. What is the average force exerted on it due to the application of its brakes? _____

- - - - - - - - - - - - - - -

3000 newtons. Kinetic energy of the moving car $= \frac{1}{2} mv^2$

$$KE = \frac{1}{2} \cdot 1000 \text{ kg} \cdot (30 \text{ m/s})^2$$
$$= 450{,}000 \text{ joules}$$

This much energy must be dissipated by exerting some force through the distance used to stop the car, 50 meters in this case.

$$\text{Work} = \text{Force} \cdot \text{Distance}$$
$$450{,}000 \text{ J} = F \cdot 150 \text{ m}$$
$$F = 3000 \text{ newtons}$$

MACHINES

1.   Suppose you push down on the level in the figure below. From experience, how much force would you need to lift the 100 pound weight—100 pounds, more, or less?

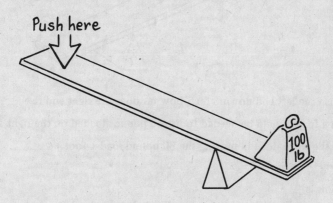

less than 100 pounds

20.   Is the law of conservation of energy violated in the figure in frame 19? No, because we have considered only force here, while work and energy involve both force and distance. If there is no friction, the product of "the force applied at the left end" and "the distance the force moves" equals the product of "100 pounds" and "its distance moved."

The preceding argument applies to all machines, including levers, pulleys, inclined planes, wheels and axles, and any combination of these. No more energy can be extracted than is put into a machine.

(a)   In the example above, in order to raise the 100 lb weight one foot, you would have

to apply a force of 25 pounds through what distance? _____

(b)   How does the lever conserve energy? _____

_____

(a)   4 ft
(b)   The product of the force applied and the distance that force moves is equal to the product of the load lifted and the distance the force moves. Thus work input equals work output.

21.   Consider a lever with dimensions given in the figure below. As the lever is turned on the pivot, the ratio of the distances traveled by the ends is equal to the ratio of their distances from the pivot point.

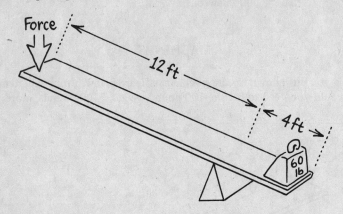

(a)   If you push the left end down 3 feet, how far does the right end rise? _____

(b)   How much force would be needed to lift 60 pounds loaded on the right side? (Hint: Consider the work done in moving the 60 pound load 1 foot.) _____

— — — — — — — — — — — — — — — — —

(a)   1 foot. This is $\frac{4}{12} \cdot 3$
(b)   20 lb. Solution: Work done by force = Work done on load
$$F \cdot 3\ \text{ft} = 60\ \text{lb} \cdot 1\ \text{ft}$$
$$F = 20\ \text{lb}$$
Notice that the ratio of forces is equal to the inverse ratio of the distances from the pivot.

SELF-TEST

The questions below will test your understanding of this chapter. Use a separate sheet of paper for your diagrams or calculations. Compare your answer with the answers provided following the test.

1.   How is work defined? _____

_____

2.   How much work is done in *holding* a piano at shoulder height for 15 minutes if the piano weighs 900 pounds and your shoulder is 5 feet above ground? _____

_____

3. How is power defined? _____

_____

4. What is the energy called that an object has due to its height? _____

5. As an object falls toward the ground, what happens to its potential energy?

_____

6. Suppose two objects have the same mass, but one is moving twice as fast as the
    other. The kinetic energy of the fast one is how many times as great as the kinetic

    energy of the slow one? _____

7. How can a machine increase the energy output of an engine? _____

_____

8. What happens to the kinetic energy of an object as the object falls? _____

_____

9. Suppose two objects are at rest on a table with a firecracker between them. The fire-
    cracker explodes, causing the two objects to move apart rapidly. Since these objects
    then have energy they did not have before, it seems that the law of conservation of

    energy has been violated. Explain. _____

_____

10. Lifting a 5 kg object 3 meters requires how much energy? _____

11. If the object in question 10 is lifted in 2 seconds, what is the power of the lifter?

    _____

12. A 3 horsepower engine can (theoretically) lift 550 pounds how far in 2 seconds?

    _____

13. A 6 pound object is 4 feet above the floor. As it falls, it is caused to lift a 2 pound
    object. What is the maximum height to which the 2 pound object can be lifted?

    _____

14. *Optional.* A 2000 kg truck is stopped by a force of 1500 newtons in a distance of

    600 meters. What was the speed of the truck? _____

15. *Optional.* An object is going 10 m/s when it hits the ground. From how high did it
    fall? (Hint: Since the object's mass does not matter, you may choose any mass you

    want.) _____

Answers to Self-Test

If your answers do not agree with those given below, review the frames indicated in parentheses before you go on to the next chapter.

1. the product of the force exerted and the distance through which it is exerted (frame 1)

2. No work is done.   (frame 1)

3. Power is work done divided by the time used in doing the work.   (frame 3)

4. potential energy   (frame 6)

5. decreases   (frame 6)

6. four times (from the formula $KE = \frac{1}{2} mv^2$)   (frame 12)

7. It can't! This would violate the law of conservation of energy.   (frame 20)

8. increases   (frames 11–12)

9. The chemical energy within the firecracker has been converted to kinetic energy of the objects. While it is true that the *kinetic* energy has increased, another energy decreased (chemical energy). Heat, sound, and kinetic energy were produced. (frame 14)

10. 147 joules (Remember, you must use the *weight* of the object, and the weight is equal to the mass times the acceleration of gravity.)   (frame 10)

11. 73.5 watts (147 J/2 s = 73.5 J/s)   (frame 5)

12. 6 feet (Its power is 1650 ft·lb/s. Thus it can do 3300 ft·lb of work in 2 seconds. This corresponds to lifting 550 pounds 6 feet.)   (frame 4)

13. 12 feet (energy input = energy output)   (frame 20)

14. 30 m/s (Calculate the work done—force × distance—and set this equal to the kinetic energy—$\frac{1}{2} mv^2$. Solve for v.)   (frame 18)

15. 5.1 meters ($mgh = \frac{1}{2} mv^2$, thus $gh = \frac{1}{2} v^2$, and $h = v^2/2\,g$.)   (frames 15–16)

CHAPTER THREE
Momentum is Conserved, Too

Prerequisites: Chapters 1 and 2

Although energy is perhaps the single most important concept in physics (other than the fundamental quantities of mass, length, time, and electric charge), another conservation law must be understood to fully predict what will happen in many situations involving force and motion. This is, as the title suggests, the conservation of momentum.

After a study of this chapter, you should be able to:

- define momentum and state its metric units;

- relate momentum to impulse;

- given a change in momentum and the time involved, calculate the average force used;

- explain how momentum is conserved in an isolated system;

- apply the principle of conservation of momentum to analysis of simple collisions;

- calculate velocity after a collision, given initial velocity and masses of the objects;

- recognize the part played by the earth in conservation of momentum.

DEFINING MOMENTUM

1.   Momentum is defined as the product of the mass of an object and its velocity. Thus the SI unit of momentum is kg·m/s. What two quantities could you change to change the momentum of an object? _____

— — — — — — — — — — — — — — — —

the object's mass or its velocity

2.   The definition of acceleration is the change in velocity divided by the time involved, or:

$$a = \frac{v_1 - v_0}{t}$$

Now recall Newton's second law in equation form:

$$F = ma$$

If we substitute for a the expression written above, we get:

$$F = m \; \frac{v_1 - v_0}{t}$$

$$Ft = m(v_1 - v_0)$$

On the right side of this equation we have the product of mass and change in velocity. This product is the change in momentum.

(a)   What are the units for the terms of the left side of the equation? _____

(b)   What are the units for the right side of the equation? _____

(c)   Show that the units on the two sides are basically the same.

_____

_ _ _ _ _ _ _ _ _ _ _ _ _ _ _ _

(a)   newton-second (or N·s)
(b)   kilogram-meter/second (or kg·m/s)
(c)   N·s = (kg·m/s^2) · s   (a newton was defined as a kg·m/s^2)
          = kg·m/s   (which is what we had on the right side)

IMPULSE

3.   The product of force and time is defined as *impulse*. In the last frame we derived the equation $Ft = m(v_1 - v_0)$.

(a)   What is the relationship between impulse and change in momentum? _____

_____

(b)   A 0.150 kilogram baseball comes toward the bat at a speed of 30 meters per second. The batter swings and sends the ball back in the direction from which it came at a speed of 40 m/s. By how much did the ball change its momentum?

_____

(c)   If the bat was in contact with the ball for 0.010 second, what average force did it

exert on the ball? _____

_ _ _ _ _ _ _ _ _ _ _ _ _ _ _ _

(a)   they are equal
(b)   10.5 kg·m/s. Solution: $m(v_1 - v_0) = .150$ kg [30 m/s − (−40 m/s)]
                                    $= .150$ kg [70 m/s]
                                    $= 10.5$ kg·m/s
      (Notice that the change in direction of the ball resulted in a velocity change of 70 m/s.)

(c)   1050 newtons.   Solution:

$$Ft = 10.5 \text{ kg} \cdot \text{m/s}$$
$$F(0.010 \text{ s}) = 10.5 \text{ kg} \cdot \text{m/s}$$
$$F = 1050 \text{ kg} \cdot \text{m/s}^2$$
$$F = 1050 \text{ newtons}$$

MOMENTUM CONSERVATION

4.   Consider the two carts in the figure below. There is a compressed spring between them, but a rope is holding them together. Since the carts are in equilibrium, the total force on each of them is zero. If the rope is cut, however, there is an unbalanced force on each of the objects by the compressed spring.

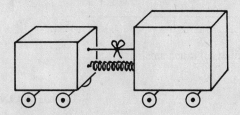

What does Newton's third law (the action-reaction law) tell us about the forces on

the carts? _____

- - - - - - - - - - - - - - - -

They are equal and opposite.

5.   Since these forces are exerted as long as the spring is in contact with the carts, and the force is exerted on one exactly as long as it is exerted on the other, each cart receives the same impulse.

(a)   Since impuse is equal to the change in momentum, what do you conclude about the

change in momentum of the two carts? _____

(b)   What is the total momentum of the system before the rope is cut? _____

(c)   What is the total momentum of the system while the carts are moving? _____

(d)   What can you conclude about momentum in an isolated system? _____

_____

- - - - - - - - - - - - - - -

(a)   They are equal and opposite (thus the total momentum is still zero).
(b)   zero
(c)   zero (one in a positive direction and one in a negative direction)
(d)   momentum is conserved in an isolated system ("Isolated system" means that there are no forces exerted on the system from outside. If you had pushed on the carts with your hand, they would not form an isolated system.)

6.   *The total momentum in any isolated system before any event is equal to the total momentum after the event.* When a gun fires a bullet, it can be considered an isolated

system until it recoils back against the shoulder and the shoulder exerts an outside, unbalanced force on it. Before the "event"—the firing of the bullet—the gun and bullet are at rest, so their total momentum is zero. After the bullet is fired, it has momentum. (Although its mass is fairly small, its speed is great, so its momentum is considerable). If momentum is to be conserved, something must gain momentum in the direction opposite to that of the bullet. The gun does just that.

(a)   How does the mass of the gun compare to that of the bullet? _____

(b)   After the firing, how does the momentum of the gun compare to that of the bullet?

       _____

(c)   Therefore how does the velocity of the gun compare to that of the bullet after

       firing? _____

— — — — — — — — — — — — — — —

(a) much greater; (b) equal; (c) much smaller

COLLISIONS

7.   The law of conservation of momentum also allows us to analyze collisions between objects. Consider the simple case of one pool ball colliding head-on with another at rest. Before the collision, the total momentum of the system is the product of the mass of the moving ball and its velocity. Those of you who have played pool know what happens after such a collision: the ball which had been moving stops, and the other ball moves forward. The conservation of momentum can tell us what the speed of that second ball must be.

(a)   Assuming their masses to be the same, what will be the momentum of the second

       ball after collision compared to that of the first ball before collision? _____

(b)   What will be the velocity of the second ball compared to what the first's was?

       _____

— — — — — — — — — — — — — — —

(a) equal; (b) equal

8.   The collision discussed above is the simplest type. Not only was the momentum after collision equal to the momentum before, but the kinetic energy of the system was the same before and after the event. Such a collision is called *perfectly elastic*. In general, collisions are not perfectly elastic; there is less kinetic energy after a collision than before. (The law of conservation of energy is not violated because there is an increase in heat energy involved.) What do you think we call collisions in which there is a decrease in the

kinetic energy? _____

— — — — — — — — — — — — — — —

inelastic

9.   Consider a wreck in which a car runs into the rear of a stationary truck. Before the collision, the car has momentum forward. We will suppose that the car and truck stick together after the crash.

Before

After

The law of conservation of momentum tells us what the momentum of the wreckage will be.

(a)   What will it be equal to? _____

(b)   What values would we need to know to calculate the velocity of the wreckage after

impact?_____

_ _ _ _ _ _ _ _ _ _ _ _ _ _ _

(a)   the momentum of the car before collision
(b)   masses of car and truck, and pre-wreck velocity of the car

10.   Since momentum is the product of a vector and a scalar, momentum is also a vector. To see what effect this has, let's go back to the pool table. In Figure 3.1 (part 1), ball A is about to collide with ball B again, but it is coming in a direction such that it will not hit it straight on. You would expect the balls to continue somewhat as shown in part 2. Before the collision, the total momentum of the system was in a downward direction on the page. After collision, the momentum of ball A is not only downward, but also toward the *right*.

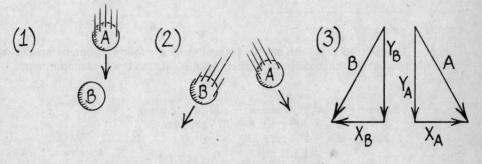

Figure 3.1

What is necessary for momentum to be conserved? _____

_____

- - - - - - - - - - - - - - -

an equal momentum toward the left

11.   In Figure 3.1 (part 3), arrow X_A, which represents the sideways momentum of A after collision, must be the same length as arrow X_B, representing ball B's sideways momentum. Vectors Y_A and Y_B represent the downward momenta (plural of momentum!) of balls A and B. The total of these two vectors must equal what momentum?

_____

- - - - - - - - - - - - - - -

the momentum of A before collision

SIMPLE MOMENTUM PROBLEMS

12.   A gun has a mass of 10 kilograms and its bullet has a mass of 20 grams. Suppose that the bullet fires out of the gun barrel with a speed of 300 m/s (something like 600 miles per hour). What is the recoil speed of the gun? (Hint: Calculate the momentum of the

bullet first.) _____

- - - - - - - - - - - - - - -

0.6 m/s.   Solution: ("G" stands for "gun" and "b" for "bullet":)
$$m_G v_G + m_b v_b = 0$$
$$10 \text{ kg} \cdot v_G + .02 \text{ kg}^* \cdot 300 \text{ m/s} = 0$$
$$v_G = -0.6 \text{ m/s}$$
(The negative answer tells us that the velocity is opposite in direction to the 300 m/s.)

13.   A ball of mass 200 grams (a softball, perhaps) is moving at 3 m/s before it collides head-on with a baseball of mass 150 grams. After collision the baseball is moving at a speed of 2.5 m/s. How fast is the softball moving after collision? (Hint: You may use grams as your unit of mass here as long as you use it on both sides of the equation.)

_____

- - - - - - - - - - - - - - -

1.125 m/s (If you failed to get this answer, try again with the following hint before checking the solution below: the momentum after collision is equal to the total momenta of the two balls.)
  Solution: $mv_{before} = mv_{after}$
$$200 \text{ gm} \cdot 3 \text{ m/s} = 200 \text{ gm} \cdot v + 150 \text{ gm} \cdot 2.5 \text{ m/s}$$
$$600 \text{ gm} \cdot \text{m/s} = 200 \text{ gm} \cdot v + 375 \text{ gm} \cdot \text{m/s}$$
$$225 \text{ gm} \cdot \text{m/s} = 200 \text{ gm} \cdot v$$
$$v = 1.125 \text{ m/s}$$

_____

*Note that mass must be expressed in the same units on each side of the equation. You can't use "20 gm" unless you converted 10 kg to grams.

Before

14.   The situation in this problem is similar to that of the two colliding pool balls of frame 10, but this time Fast Eddie has put in an extra-massive ball—it has three times as much mass as the regular ball. He shoots this ball at an unsuspecting regular ball which is at rest. They hit, and the massive ball rolls as shown in the figure. Note that Fast Eddie's ball now has a momentum toward the right.

(a)   What do we know about the direction in which the regular ball moves?

_____

(b)   How does the speed (across the page) of the regular ball compare to that of the massive ball?

Large ball

After

_____

(a)   downward and toward the left

(b)   greater (3 times as great) (Because there was no sideways—"across the page"—momentum before the collision, there can be none after. Thus since one ball is only one-third as massive, it must move across the page three times as fast.)

FRICTION AND THE EARTH

15.   We have thus far neglected to mention the effect of friction. We spoke of the momentum of cars and trucks immediately before and after collisions, for example, but said nothing of the fact that after a wreck the wreckage would slide to a stop. What happens to the momentum of the wreckage? The answer is that the momentum is transferred to the earth. The earth spins a little faster (or slower, depending upon whether the motion was westward or not). It may at first glance seem hard to believe that an object as small as a truck can affect the speed of the earth, but one must remember that we are not saying that the effect would be great enough to measure. Since the mass of the earth is so tremendous, its change in speed would be very, very small. Undetectable—but there.

Consider the recoiling gun when it hits the shooter's shoulder. The system is no longer "isolated," but includes the entire earth. Is the change in momentum of the

shooter detectable? _____

_____

Perhaps (he may fall over backwards). But if the shooter is ready, he braces against the expected impulse and the momentum is transferred to the earth.

SELF-TEST

The questions below will test your understanding of this chapter. Use a separate sheet of paper for your diagrams or calculations. Compare your answers with the answers provided following the test.

1. How are impulse and momentum related? _____

    _____

2. How does the momentum of a bullet as it leaves the barrel of a gun compare to the

    recoil momentum of the gun? _____

    _____

3. A balloon is blown up and then released. It "zips" forward, and thus has momentum which it did not have before. Does this situation violate the law of conservation

    of momentum? If not, explain how momentum is conserved. _____

    _____

4. Suppose a moving object collides with a stationary one and the two stick together. How will the speed of the stuck-together objects compare to the speed of the origi-

    nal moving object? _____

5. How would the result of a perfectly elastic collision differ from that of an elastic

    collision? _____

    _____

6. Suppose you drop a ball and it rebounds with less speed than it hit. Is the collision

    with the floor elastic or inelastic? _____

7. Is the law of conservation of momentum violated if a gun is attached so firmly to a

    support that it cannot recoil? Explain. _____

    _____

8. Is momentum a scalar or a vector quantity? _____

9. If the total momentum of a system before a collision can be represented by an arrow pointed toward the right, what can be said of the direction of the total

    momentum after collision? _____

    _____

10. Two blocks of ice are sitting on a table with a piece of dynamite between them. One block of ice has a mass of 12 kg and the other's mass is 18 kg. The dynamite explodes. If the block of lesser mass moves off with a speed of 10 m/s, what is the

    speed of the more massive block? _____

11.   A 2000 pound car moving at 50 miles per hour collides with a stationary car weighing 3000 pounds. If the two cars stick together, what is their speed immediately after collision? (Hint: You may use "pounds" for mass and "miles per hour" for velocity, as long as you do so on both sides of the equation. This is permissible because weight (pounds) is proportional to mass, and miles/hour is a legitimate unit of velocity.) _____

Answers to Self-Test

If your answers do not agree with those given below, review the frames indicated in parentheses before you go on to the next chapter.

1.   The impulse applied to an object is equal to the change in momentum of the object. (frame 3)

2.   The two momenta are equal (but opposite in direction).   (frame 6)

3.   No, the law of conservation is not violated. The air forced out of the balloon is given a momentum equal to the balloon (but opposite in direction, naturally). (frame 6)

4.   It will be less. (The momentum of the stuck-together objects must be equal to the momentum of the original moving object. Since the two objects have a mass greater than the mass of the single object, the speed of the two objects must be less. (frame 9)

5.   In a perfectly elastic collision, kinetic energy is conserved, but in an inelastic collision it is not.   (frame 8)

6.   inelastic   (frame 8)

7.   No. The support—and the entire earth—recoils.   (frame 15)

8.   vector   (frame 10)

9.   Again it can be represented by an arrow pointing to the right (and of the same length as the first arrow).   (frame 11)

10.   6.67 m/s. Solution: The total momentum before the explosion is zero, so the momentum of one piece of ice after explosion plus the momentum of the other piece must equal zero.

$$12 \text{ kg} \cdot 10 \text{ m/s} - 18 \text{ kg} \cdot v = 0$$
$$v = -6.67 \text{ m/s}$$

(The minus sign tells us that the speed of this block is in the opposite direction from the other block.)   (frame 12)

11.   20 miles/hour. Solution:

$$(2000 \text{ lb} \cdot 50 \text{ mi/hr}) + (3000 \text{ lb} \cdot 0) = 5000 \text{ lb} \cdot v$$
$$100,000 \text{ lb} \cdot \text{mi/hr} = 5000 \text{ lb} \cdot v$$
$$v = 20 \text{ mi/hr} \quad \text{(frame 13)}$$

CHAPTER FOUR

Gravity

Prerequisite: Chapter 1

The law of gravitation plays a very prominent role in our view of the universe. It is the force of gravity which causes a pin to fall to the floor, which holds our planet earth in its orbit around the sun, which holds the sun and some 200 billion other stars in our Milky Way galaxy, and which causes the gigantic galaxies to cluster together in groups. Yet the law of gravity was unknown to humankind until rather recently in history. And here again the name of Isaac Newton appears; he is the genius who first hypothesized that the heavens and the earth are united by this single, all-encompassing principle.

A study of this chapter should enable you to:

- state the law of gravity using words or a formula;

- explain how gravity keeps the moon in orbit;

- state the shape of a planet's orbit and the sun's position with respect to that orbit;

- give examples to illustrate the effect of distance on the gravitational force between two objects;

- explain how the law of equal areas defines the speed of a planet around the sun;

- specify the effect on orbiting satellites of the force of gravity;

- justify the concept of "apparent weightlessness";

- use numerical examples to illustrate how the acceleration of gravity varies with height above the earth;

- describe the relationship between Kepler's laws and Newton's laws;

- give the value of the universal gravitational constant (optional);

- calculate gravitational attraction between two objects, given masses and distances (optional);

- use the law of gravity to calculate the mass of the earth (optional).

THE LAW OF GRAVITY

1.   According to Newton's law of gravity, *every object in the universe attracts every other object in the universe with a force proportional to the objects' masses and inversely proportional to the square of the distance between their centers.* The statement is much shorter in symbolic form:

$$F \sim \frac{m_1 \cdot m_2}{d^2}$$

The symbol $\sim$ means "is proportional to"; m_1 symbolizes the mass of one of the objects, and m_2 is the mass of the other; d is the distance between the objects.

The law says that there is a force of gravity between *every* two objects, between you and the person next to you, and between you and a bus out on the street. Why do you not feel the force of gravity between you and this book? _____

_____

– – – – – – – – – – – – – – – –

The masses involved are so small that the force is very small.

2.   Consider how the force of gravity between two objects changes as the distance between them changes. An *inverse square* relationship is involved—the force changes inversely as the square of the distance. Thus if the distance between two objects is tripled, the force becomes 1/9 as great. ($1/3^2 = 1/9$)

(a)   If the distance is made four times as great, how does the force of gravity change?

_____

(b)   Suppose two objects are initially 5 meters apart and are then moved to 1 meter

apart. How does the force change? _____

– – – – – – – – – – – – – – – –

(b) becomes $\frac{1}{16}$ as great; (b) 25 times greater

3.   The distance involved in the law of gravity is the distance between the centers of the two masses. Our distance to the center of the earth is about 4000 miles when we are on the surface.

(a)   If we go to a height of 4000 miles above the surface of the earth, how does the

force of gravity (our weight) change? _____

(b)   Suppose you weigh 117 pounds here on earth. How far from the center of the earth

must you go to weigh 13 pounds? _____

– – – – – – – – – – – – – – – –

(a)   It becomes $\frac{1}{4}$ as much. (You are then 8000 miles—twice as far—from center.)
(b)   12,000 miles ($117 \div 13 = 9$. To decrease the force to $\frac{1}{9}$, you must be 3 times as far away.)

4.   In Chapter 1 you learned that the acceleration of gravity is 32 ft/s^2 or 9.8 m/s^2. This is not true, however, for locations at different distances from the center of the earth; at different distances the force of gravity is different. On the surface of the earth this difference is slight, the acceleration due to the force of gravity ranging from about 9.788 m/s^2 to about 9.808 m/s^2. In Chapter 1, frame 11 we found that in the first second of fall, an object falls 16 feet. This approximation is not strictly true at all locations on earth.

What causes the acceleration due to gravity to be different at different distances?

_____

– – – – – – – – – – – – – – – –

The force of gravity is different.

5.   Since the distance fallen from rest varies directly with the object's acceleration ($d = \frac{1}{2}at^2$—Chapter 1, frame 11), and the acceleration varies directly as the net force applied to the object, the distance fallen in one second is directly proportional to the force of gravity. Thus if the force of gravity is 1/4 as great, an object falls 4 feet instead of 16 feet during the first second of fall.

(a)   As one rises above the surface of earth, does the acceleration of gravity become

greater or less? _____

(b)   Does an object fall more or less far during the first second of fall if it is high above

the earth compared to sea level? _____

(c)   Suppose an object is 60 times as far from earth as the earth's surface is from earth-

center. What fraction of 16 feet will the object fall in one second? _____

– – – – – – – – – – – – – – – –

(a) less; (b) less far; (c) 1/60^2, or 1/3600

THE MOON AND GRAVITY

6.   The answer to the last question in frame 5 applies to any object, including the moon (which, in fact, is 240,000 miles from earth-center, or 60 times as far as we are from center). Consider the moon moving in a direction shown by the solid arrow in the figure. The moon should continue in a straight line if no force is applied to it. In fact, a force of gravity is exerted on it by the earth, so instead of going in a straight line, its path is bent by the force. In one second, the moon "falls" from a straight-line path by just the amount predicted by Newton's law of gravity, 1/60^2 of 16 feet. If you do the calculations, you will find that this distance is 0.00444 ft, or 0.0533 in. Isaac Newton did this calculation and realized that it confirmed his law of gravity. It was the first application of a single law of nature to both heavenly objects and earthly ones.

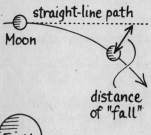

(a)   Which of Newton's laws says that the moon

should go in a straight line? _____

(b)   Which explains why it doesn't? _____

— — — — — — — — — — — — — — — —

(a) Newton's first law (Chapter 1, frame 13); (b) Newton's law of gravity

KEPLER'S LAWS

7.   About 50 years before Newton formulated the law of gravity, Johannes Kepler
(1571–1630) had found some rules which could be used to calculate the orbits of the
planets. The reason that these rules worked was not known until Newton applied his
gravitational theory to planets, however. We will consider Kepler's laws and see how they
are explained by the laws of Newton.

According to Kepler's first law, planets orbit the sun in the shape of an ellipse. An
ellipse is the shape that a circle appears when one looks at it from an angle. The easiest
way to draw an ellipse is to put two tacks in a piece of paper and put a loose loop of
string around them as shown in the first part of the figure below. Now take a pencil and
stretch the string out. Then sweep around the tacks, always keeping the string taut. The
figure drawn is an ellipse, and the two points where the tacks are located are called the
foci (the plural of *focus*) of the ellipse. Kepler's first law further states that the sun is
located at one focus of the elliptical path of the planet.

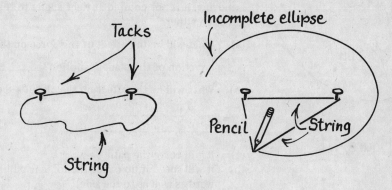

(a)   If the tacks are placed farther apart, how will the shape of the ellipse be changed?

_____

(b)   What would the positions of the tacks have to be to draw a circle? _____

_____

(c)   State Kepler's first law. _____

_____

— — — — — — — — — — — — — — — —

(a)   it will be flatter
(b)   both at the same point (Thus a circle is just a special form of an ellipse.)
(c)   Planets orbit the sun in elliptical paths with the sun at one focus of the ellipse.

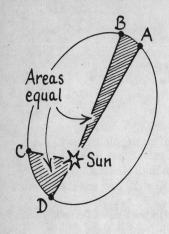

8.   As a planet orbits the sun, a line from the planet to the sun sweeps across equal areas in equal times. To see what this means, refer to the figure. Suppose that the earth moves from point A to point B in a two-week period. Later in the year, it is closer to the sun at point C. In the next two weeks it will go just far enough that the area swept out by an imaginary line from the sun will be equal to the area it swept out when it moved from A to B. Does the planet move faster between A and B or between C and D?

_____

_ _ _ _ _ _ _ _ _ _ _ _ _ _ _ _

between C and D

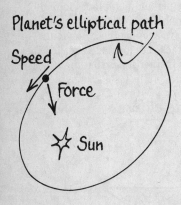

9.   Now we'll see how Newton's laws explain this apparently arbitrary rule found by Kepler. In Figure 4.1, the force of the sun's gravity on the planet is represented by the arrow. Note that although the arrow is not pointed in the direction the planet is moving, it is not pointed at right angles to that direction either.

(a)   What will be the effect of this force on the direction of the planet's motion? _____

(b)   What will be the effect on the planet's speed?

_____

_ _ _ _ _ _ _ _ _ _ _ _ _ _ _ _

Figure 4.1

(a)   it will curve the path
(b)   it will speed it up (since the force is pulling forward as well as to the side)

10.   Next consider the motion of the planet when it is halfway around its orbit from where it is shown in Figure 4.1. What will be the effect of the force here?

_____

_ _ _ _ _ _ _ _ _ _ _ _ _ _ _ _

It will still curve the path, but now it will slow down the planet.

11.   If you do the necessary math, you will find that the amount of speedup or slowdown of the planet would be just the right amount to give the results which Kepler stated as his law of equal areas. This is a very important point: Kepler's laws were simply concise statements of how planets moved, but they did not relate planet motion to any other law

of nature. Newton's laws apply to objects here on the surface of the earth as well as in space.*

(a)   What is the shape of the earth's orbit around the sun? _____

(b)   Does the earth move, at a constant speed around the sun? _____

(c)   What force causes the shape and speed of the earth's orbit? _____

– – – – – – – – – – – – – – – –

(a) ellipse; (b) no; (c) gravity

EARTH SATELLITES

12.   The figure below is similar to one found in Newton's book, *The Principia.* Newton argued that if a cannon could be placed on a hill high enough to be outside the earth's atmosphere, and if it could be made powerful enough, it could place a cannonball in orbit around the earth.

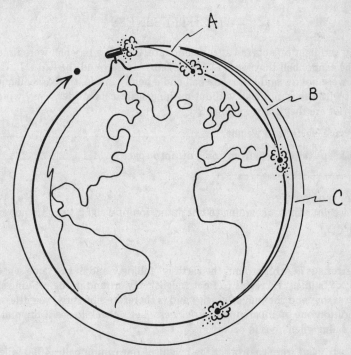

First, suppose that it is just an ordinary cannon on this high mountain. In that case the path of the cannonball might be like path A in the figure. If it were more powerful, it might take a path like B, or even C. And finally, if such a cannon could be obtained, it could shoot the cannonball so that it came back to where it started. It would then be in

*We have neglected the third law of Kepler. It relates the distance from a planet to the sun to the time required for the planet to travel around the sun. It likewise is explained by Newton's laws.

orbit. What keeps such a cannonball from flying into space? _____

— — — — — — — — — — — — — — —

the force of gravity

13.   Consider that the cannon is 4000 miles above the earth's surface.

(a)   How far will the ball fall in the first second (due to gravity)? _____

(b)   Why do you suppose the cannonball never reaches the ground? _____

_____

— — — — — — — — — — — — — —

(a)   4 ft ($1/2^2$ · 16 ft = 4 ft, see frame 5)
(b)   The curvature of the earth is just enough so that as the cannonball falls, it gets no closer to the surface.

WEIGHTLESSNESS

14.   Heavy and light objects fall at the same rate. Thus if a baseball were fired side-by-side with the cannonball they would "fall" side-by-side around the earth. In fact, if the cannonball were hollow and had a baseball and a penny inside, the baseball and penny would orbit with the cannonball without pushing against its insides. They would seem *weightless*. But are they truly weighless?

(a)   How have we defined weight? _____

(b)   What keeps the objects from going off into space? _____

— — — — — — — — — — — — — —

(a)   the force of gravity
(b)   their weight (Thus, according to our definition of weight, they are not truly weightless.)

15.   An astronaut in orbit around the earth is "falling," and in the same sense that the cannonball is "falling." He seems to float weightlessly around in the orbiting ship. He is now, however, beyond the pull of gravity and is therefore not truly weightless. We should call his condition one of *apparent weightlessness*. Let's check to see if the pull of gravity is far less on him when he is in orbit.

(a)   If an astronaut were to orbit at the height of our cannonball—4000 miles—what would be his relative weight? _____

(b)   Normally our manned orbiters are only a few hundred miles above the surface. In such nearby orbits, the pull of gravity is still 80 to 90 percent of what it is on the surface. Is the astronaut's weight closer to what it is on earth or to the 4000 mile figure? _____

— — — — — — — — — — — — — —

(a)   1/4 of his "normal" weight (as in frame 3)
(b)   much closer to his earth weight

CALCULATIONS

16. Earlier we stated the law of gravity in symbols as:

$$F \sim \frac{m_1 \cdot m_2}{d^2}$$

In order to use this relationship for solving problems, we must remove the proportionality by placing a constant in the expression:

$$F = G \cdot \frac{m_1 \cdot m_2}{d^2}$$

The constant G is known as the *universal gravitational constant*. The value of G is:

$$G = 6.67 \cdot 10^{-11} \frac{N \cdot m^2}{kg^2} *$$

(Force is to be measured in newtons, distance in meters, and mass in kilograms.) Calculate the gravitational attraction between two people standing 1 meter apart. Take the mass of each person as 70 kg. (This is about the mass of a 155 lb person.)

$F = $ _____

– – – – – – – – – – – – – – – – –

$3.3 \cdot 10^{-7}$ newtons. Solution:

$$F = 6.67 \cdot 10^{-11} \frac{N \cdot m^2}{kg^2} \cdot \frac{70 \text{ kg} \cdot 70 \text{ kg}}{(1 \text{ m})^2}$$

Recalling that a newton is about one-fourth of a pound, we see that this is less than 10^{-7} lb, or less than one ten-millionth of a pound! No wonder we do not experience it!

17. Historically, the gravitational force was carefully measured between two "ordinary" objects with masses such as in the last frame. This allowed a calculation of the constant G. Once G was known, the mass of the earth—until then unknown—was calculated. You can do it, too.

On the surface of the earth we are $6.4 \cdot 10^6$ meters from the center. In Chapter 1 we learned that a 1 kilogram object weighs 9.8 newtons. To use these facts to calculate the mass of the earth, let m_1 be the earth's mass.

(a)   Solve the equation for m_1. $m_1 = $ _____

(b)   Calculate m_1, the mass of the earth. $m_1 = $ _____

– – – – – – – – – – – – – – – –

(a) $m_1 = \dfrac{F \cdot d^2}{m_2 \cdot G}$ (F is the weight of the object with mass m_2.)

(b) $m_1 = 6 \cdot 10^{24}$ kg. Solution: $m_1 = \dfrac{9.8 \text{ N } (6.4 \cdot 10^6)^2}{1 \text{ kg} \cdot 6.67 \cdot 10^{-11} \text{ N} \cdot m^2/kg^2}$

_____

*See Appendix I for an explanation of powers of ten notation.

SELF-TEST

The questions below will test your understanding of this chapter. Use a separate sheet of paper for your diagrams or calculations. Compare your answer with the answers provided following the test.

1. State the law of gravity, including the dependence of the force on the masses and distance involved. _____

    _____

2. How far must a person rise above the *surface* of the earth to decrease his or her weight to 1/4 the amount it is on the surface? To 1/25th of his or her surface weight? _____

3. If the earth were twice as massive, but still the same size, how would our weight differ? _____

4. What is Kepler's first law (the law of ellipses)? _____

    _____

5. Using Newton's laws of motion and gravity, explain the fact that the moon's path continually curves toward the earth. _____

    _____

6. Both Kepler's and Newton's laws accurately described the paths of the planets. What advantage did Newton's laws have? _____

    _____

7. Use Newton's laws to explain why a planet speeds up during part of its orbit.

    _____

    _____

8. Are astronauts in orbit truly weightless? Explain. _____

    _____

9. *Optional.* What gravitational force is exerted between two 1 kilogram masses one meter apart? Two meters apart? _____

10. *Optional.* The mass of the moon is about $7.5 \cdot 10^{22}$ kg, and its *diameter* is about $3.5 \cdot 10^6$ meters. Calculate the weight of a 70 kg person on the moon. (On earth a 70 kg person weighs about 700 newtons, or 155 pounds.) _____

    _____

Answers to Self-Test

If your answers do not agree with those given below, review the frames indicated in parentheses before you go on to the next chapter.

1.   Every object in the universe attracts every other object with a force proportional to the objects' masses and inversely proportional to the square of the distances between their centers. Or:

$$F \sim \frac{m_1 \cdot m_2}{d^2} \quad \text{(frame 1)}$$

2.   4000 miles, 16,000 miles   Solution: Taking the radius of the earth as 4000 miles, one must get twice as far from its center to decrease one's weight to 1/4 the surface value. This is 4000 miles above the surface. To decrease weight to 1/25, one must move 5 times as far away as the surface. This is 20,000 miles from center, or 16,000 miles from the surface.   (frame 3)

3.   twice as much   (frame 1)

4.   A planet orbits the sun with its path being the shape of an ellipse with the sun at one focus of the ellipse.   (frame 7)

5.   If there were no unbalanced forces acting on the moon, it would continue in a straight line. The force of gravity acts to change its direction of motion toward the earth.   (frame 6)

6.   Newton's laws also applied here on earth, while Kepler's laws seemed to have no relation to motions on earth.   (frame 11)

7.   As the planet is moving along its ellipse in a direction toward the sun, the effect of the force of gravity is to speed up the planet as well as curve its path.   (frame 9)

8.   Not if weight is defined as the force of gravity. The astronauts are in free fall along with their spaceship. They fall together the same as two coins fall side-by-side when dropped.   (frames 14–15)

9. $6.67 \cdot 10^{-11}$ N, $1.67 \cdot 10^{-11}$ N. Solution:

$$F = G \cdot \frac{m_1 \cdot m_2}{d_2}$$

$$= 6.67 \cdot 10^{-11} \frac{\text{N} \cdot \text{m}^2}{\text{kg}^2} \cdot \frac{1\ \text{kg} \cdot 1\ \text{kg}}{(1\ \text{m})^2}$$

$$= 6.67 \cdot 10^{-11}$$

At 2 meters distance, the force will be one-fourth of this, or $1.67 \cdot 10^{-11}$ N. (frame 16)

10.   114 N. (Substitute into the gravitational equation, being careful to use the radius of the moon, rather than the diameter, which was given.   (frame 16)

PART TWO

The Physics of Molecules and Heat

As the title of this second part of the book indicates, a primary subject matter will be that of thermal energy—"heat" in everyday language. But we will find that thermal energy in an object is actually the result of motion of the object's atoms and molecules. Thus we start the second part of the book with a study of atoms and molecules. This is followed by two chapters in which we consider the conditions of atoms and molecules in solids, liquids, and gases and how the behavior of these tiny particles explains the properties of the three states of matter. Then finally we will study the effect of heat on matter.

The first chapter in this secion, on atoms and molecules, is prerequisite for many later chapters in the book. Although the remaining chapters of the section are not a prerequisite for following chapters, their subject matter is important to physics because it forms part of the story of the flow of energy through our universe.

CHAPTER FIVE

Atoms and Molecules

No Prerequisites

Ancient man must have looked at the countless substances in his everyday experience and wondered if they were perhaps made up of a smaller number of more fundamental, elementary substances. Today we know that this is indeed true—the entire physical world is made up of a relatively few substances, and these exist as atoms and molecules. The basic nature of atoms and molecules affects almost every aspect of physics. In this chapter we take a trip to the unimaginably small world of atoms.

When you complete your study of this chapter, you will be able to:

- differentiate between elements and compounds;

- specify the total number of elements now known;

- differentiate between atoms and molecules;

- relate atoms and molecules to elements and compounds;

- give an example of the application of the Law of Definite Proportions;

- identify the relative number and type of atoms in a compound, given its formula and the periodic table;

- label the locations of the atomic particles on a drawing of the Bohr model of the atom;

- identify the component of an atom that determines its chemical properties;

- define atomic mass and atomic number;

- interpret portions of the periodic table;

- specify, using an example, the importance of the number of electrons in the outer orbit of an atom;

- use Avogadro's number in determining kilogram atomic mass;

- given the number of grams of a compound and the periodic table, calculate the number of molecules of that compound.

1.   All of the materials and substances we experience are made up of 100 or so *elements*, which are the substances into which all matter can (in principle) be divided. Some of the elements are hydrogen, oxygen, carbon, iron, silicon, sulfur, and uranium. Table 5.1 is a complete list of the elements known today and their abbreviations.

   Each element has certain characteristic properties. For example, hydrogen is a gas at ordinary temperatures, but it liquefies at a known (lower) temperature. It has a certain density for a given temperature and pressure, and it forms certain compounds when it combines with other elements. Any material which, when tested, has all the properties of hydrogen must be hydrogen. If any properties are different, it must be something other than hydrogen.

(a)   How many different elements are known today? _____

(b)   Suppose an element tests to have all the properties of iron. How likely is it to be

   iron? _____

– – – – – – – – – – – – – – – – –

(a) 106 (see Table 5.1); (b) very definitely

2.   Most substances we deal with daily are not elements but compounds—chemical combinations of two or more elements. Water, for example, is composed of hydrogen and oxygen. Carbon dioxide is composed of carbon and oxygen. Refer to Table 5.1 as needed to identify each of these as an element or compound.

(a)   water _____   (b)   alcohol _____

(c)   carbon _____   (d)   hydrogen _____

(e)   aluminum _____

– – – – – – – – – – – – – – – –

elements: c, d, and e; compounds: a and b

3.   Any compound can be broken into its constituent elements by chemical means. For example, if an electric current is passed through water, the water is separated into hydrogen and oxygen. The properties of a compound may be greatly different from the properties of the elements which make it up. You probably know enough about the properties of hydrogen, oxygen, carbon, water, and carbon dioxide to compare the compound to its elements in these cases. An even more outstanding example is common table salt, or NaCl; one of the NaCl elements is a metal, and one is a poisonous gas.

(a)   Refer back to Table 5.1. Which elements make up table salt? _____

   _____

(b)   Can table salt be changed from a compound into two elements? _____

– – – – – – – – – – – – – – – –

(a) sodium and chlorine; (b) yes

Number	Element	Symbol	Number	Element	Symbol
1.	hydrogen	H	54.	xenon	Xe
2.	helium	He	55.	cesium	Cs
3.	lithium	Li	56.	barium	Ba
4.	beryllium	Be	57.	lanthanum	La
5.	boron	B	58.	cerium	Ce
6.	carbon	C	59.	praseodynium	Pr
7.	nitrogen	N	60.	neodymium	Nd
8.	oxygen	O	61.	prometeum	Pm
9.	fluorine	F	62.	samarium	Sa
10.	neon	Ne	63.	europium	Eu
11.	sodium	Na	64.	gadolinium	Gd
12.	magnesium	Mg	65.	terbium	Tb
13.	aluminum	Al	66.	dysprosium	Dy
14.	silicon	Si	67.	holmium	Ho
15.	phosphorus	P	68.	erbium	Er
16.	sulfur	S	69.	thulium	Tm
17.	chlorine	Cl	70.	ytterbium	Yb
18.	argon	Ar	71.	lutecium	Lu
19.	potassium	K	72.	hafnium	Hf
20.	calcium	Ca	73.	tantalum	Ta
21.	scandium	Sc	74.	tungsten	W
22.	titanium	Ti	75.	rhenium	Re
23.	vanadium	V	76.	osmium	Os
24.	chromium	Cr	77.	iridium	Ir
25.	manganese	Mn	78.	platinum	Pt
26.	iron	Fe	79.	gold	Au
27.	cobalt	Co	80.	mercury	Hg
28.	nickel	Ni	81.	thallium	Tl
29.	copper	Cu	82.	lead	Pb
30.	zinc	Zn	83.	bismuth	Bi
31.	gallium	Ga	84.	polonium	Po
32.	germanium	Ge	85.	astatine	At
33.	arsenic	As	86.	radon	Rn
34.	selenium	Se	87.	francium	Fr
35.	bromine	Br	88.	radium	Ra
36.	krypton	Kr	89.	actinium	Ac
37.	rubidium	Rb	90.	thorium	Th
38.	strontium	Sr	91.	protractinium	Pa
39.	yttrium	Y	92.	uranium	U
40.	zirconium	Zr	93.	neptunium	Np
41.	niobium	Nb	94.	plutonium	Pu
42.	molybdenum	Mo	95.	americium	Am
43.	technetium	Tc	96.	cirium	Cm
44.	ruthenium	Ru	97.	berkelium	Bk
45.	rhodium	Rh	98.	californium	Cf
46.	palladium	Pd	99.	einsteinium	Es
47.	silver	Ag	100.	fermium	Fm
48.	cadmium	Cd	101.	medelevium	Md
49.	indium	In	102.	nobelium	No
50.	tin	Sn	103.	lawrencium	Lw
51.	antimony	Sb	104.	rutherfordium	Rf
52.	tellurium	Te	105.	halmium	Ha
53.	iodine	I	106.	*	

*Element 106 has been identified, but an official name has not yet been agreed upon.

Table 5.1

ATOMS AND MOLECULES

4.   Suppose you take an iron bar and start cutting it into smaller and smaller pieces. Even with an ideally sharp knife and a perfect microscope you would not be able to keep cutting up the iron indefinitely. You would finally reach the smallest piece of iron which exists; you would have an atom of iron. An atom is the smallest part of an element which retains the properties of the element. When the atom itself is broken into parts, the resulting particles no longer retain the properties of the original element. The study of what happens when an atom is broken up is the subject of nuclear physics, to be introduced in Chapters 25 through 27.

(a)   Would an atom be the smallest piece of an element or a compound?_____

(b)   When you divide an atom, do you still have the same substance? _____

– – – – – – – – – – – – – – –

·(a)   an element
(b)   no

5.   Elements always combine in certain fixed proportions when they form a compound. For example, it always happens that when hydrogen and oxygen combine to form water, the ratio of oxygen to hydrogen is 8 to 1 by weight. If 3 grams of hydrogen are used in the reaction, 24 grams of oxygen are used. This combination by definite, fixed proportion occurs in every chemical reaction. The easiest way to explain this *Law of Definite Proportions* is that when two elements combine, there is some smallest chunk of one element that unites with one, or two, or three, or more smallest chunks of the other element.
      These smallest chunks are atoms. When water is formed from hydrogen and oxygen, two atoms of hydrogen combine with each atom of oxygen. This combination of two hydrogen and one oxygen atom is called a water *molecule.* A molecule of water is written symbolically as H_2O, the subscript 2 indicating that there are two atoms of hydrogen in the molecule.

(a)   If you have 8 atoms of hydrogen, how many atoms of oxygen would you need to

       make 4 molecules of water? _____

(b)   What would you need to make 10 molecules of carbon dioxide (CO_2)?_____

       _____

– – – – – – – – – – – – – – –

(a) 4; (b) 10 carbon and 20 oxygen atoms

6.   The chemical symbol for sulfur is S, and sulfuric acid has the formula H_2SO_4. This means that a molecule of sulfuric acid contains 2 atoms of hydrogen, 1 atom of sulfur, and 4 atoms of oxygen. A chemical formula such as H_2O or H_2SO_4 is used to symbolize two different things: sometimes (as above) it refers to a single molecule of the compound, and sometimes it refers to the compound as a whole. The meaning is normally clear from the context, so there should be no confusion.

(a)   HCl is hydrochloric acid. Twenty molecules of this acid would contain how much

       hydrogen? _____ How much chlorine?_____

(b) H_2SO_4 contains how many times more oxygen atoms than hydrogen atoms?

_____

(c)   If each oxygen atom weighs 16 times as much as each hydrogen atom, how many times as much weight of oxygen than hydrogen is used to make H_2SO_4?

_____

‒ ‒ ‒ ‒ ‒ ‒ ‒ ‒ ‒ ‒ ‒ ‒ ‒ ‒ ‒

(a)   20 atoms of hydrogen and 20 atoms of chlorine
(b)   2 times as many
(c)   32 times as much weight of oxygen (This, then, is an example of the Law of Definite Proportions.)

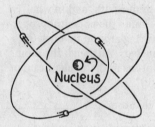

Figure 5.1

7.   You will never see an atom, not because we can't make microscopes powerful enough, but because it is simply *impossible* due to the nature of light and the properties of atoms. (We'll have more about the nature of light later.) You undoubtedly have seen drawings of atoms which make the atom appear somewhat like the solar system, with small objects revolving around a central object. (See Figure 5.1.) Such a drawing is simply a visualization of what is called the Bohr model of the atom.

(a)   Does an atom really look like Figure 5.1?_____

(b)   Do you think a microscope will someday enable you to see an atom? _____

‒ ‒ ‒ ‒ ‒ ‒ ‒ ‒ ‒ ‒ ‒ ‒ ‒ ‒ ‒

(a) no; (b) no (not a microscope using visible light, anyway)

8.   The Bohr model of the atom is a theoretical construct which helps us visualize the behavior of the atom in already-tested situations and to predict its behavior in situations

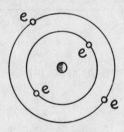

Figure 5.2

in which it has not been observed. This widely used model is named after Niels Bohr, a Danish physicist who was a leader in the development of our concept of the atom.

The Bohr atom model is made up of three particles: the proton, the electron, and the neutron. The proton has a positive electrical charge and exists in the nucleus with the electrically neutral neutrons. The negatively charged electrons revolve around the nucleus, in all directions and in all planes,* as in Figure 5.1.

_____

*The planets of the solar system all revolve in the same direction and nearly in the same plane.

(a)   If each electron contains the same amount of negative electrical charge as a proton does a positive charge, what is the resultant charge on an atom with 2 protons and 2 electrons? _____

(b)   If an atom is normally electrically neutral, what can you say about the number of electrons and protons? _____

(c)   Refer to Figure 5.2. How many protons would this atom have? _____

— — — — — — — — — — — — — — —

(a) zero, or neutral; (b) the same number of each; (c) 4, to balance the electrons

9.   Since the electrons revolve around the nucleus of an atom, they form its outer "surface" and determine what other atoms their atom will combine with, and in what ratio the combination will be made. In other words, they determine the chemical properties of the element. The number of electrons an atom has revolving around its nucleus, however, is normally equal to the number of protons in the nucleus. Although the atom may momentarily lose or gain an electron,* the number of protons in the nucleus cannot be changed by ordinary means. (A nuclear reaction is required.)

(a)   Which particles in the atom interact with those of another atom? _____

(b)   To determine whether a particular atom is of a different element than another, what do you need to know about the atoms? _____

— — — — — — — — — — — — — — —

(a) the electrons; (b) the number of protons in their nuclei

10.   Refer back to Table 5.1 and note that each element has a number to the left of its name. This number is the number of protons in the nucleus of an atom of that element. Because of the importance of this number, it is given a name, the *atomic number.*

(a)   What is the atomic number of uranium? _____

(b)   What element has only one proton in its nucleus? _____

(c)   What does the atomic number tell us? _____

(d)   How many electrons does lithium have when it is neutral? _____

— — — — — — — — — — — — — — —

(a) 92; (b) hydrogen; (c) the number of protons in the nucleus of the atom; (d) 3, because it has 3 protons

---

*Such an atom, with a fewer- or a greater-than-normal number of electrons is called an *ion.*

THE PERIODIC TABLE

11. Table 5.2 is a table of the elements in which they are arranged by chemical properties. They are listed horizontally in order by atomic number. Elements with similar
properties have been placed in the same vertical column. (The properties of elements in
the same column are *similar*, but not the same.) The most conspicuous example of similarity of properties is found in the last column on the right. These elements are called the
noble (or inert) gases, because none of them reacts readily with another element. This
table is called the *Periodic Table*. To familiarize yourself with the table, use it to answer
these questions.

(a)   Which number in each box is the atomic number?_____

(b)   What is the symbol for element number 44?_____

(c)   Is oxygen more like sulfur or hydrogen in its properties? _____

(d)   Name two noble gases. _____

– – – – – – – – – – – – – – – –

(a) the one at the top; (b) Ru, ruthenium; (c) sulfur; (d) helium, neon, argon, krypton,
xenon, radon (any two)

ELECTRON ORBITS

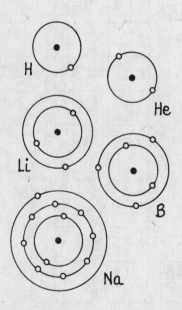

Figure 5.3

12.   Figure 5.2 showed the Bohr model with two
electron orbits, but with more than one electron in
each orbit. In the Bohr model, a number of electrons
will normally occupy a single orbit (or level), but the
maximum number which can be at any level is strictly
limited (2 in the first, 8 in the second, 8 in the third,
and so on.)* For example, helium has two electrons
orbiting at the first level; lithium has the third electron in a higher orbit. See Figure 5.3 for the following
questions.

(a)   Boron (B) has five electrons. How many will be
in its outer orbit? _____

(b)   How many electrons does sodium (Na) have all
together? _____

In the inner orbit? _____

In the second orbit?_____

In the outer orbit? _____

– – – – – – – – – – – – – – –

(a) 3; (b) 11, 2, 8, 1

*This sytem of limiting the number of electrons in an orbit and *excluding* others is
called the Pauli Exclusion Principle.

1A	2A	3B	4B	5B	6B	7B	8B			1B	2B	3A	4A	5A	6A	7A	8A
1 H 1.0																	2 He 4.0
3 Li 6.9	4 Be 9.0											5 B 10.8	6 C 12.0	7 N 14.0	8 O 16.0	9 F 19.0	10 Ne 20.2
11 Na 23.0	12 Mg 24.3											13 Al 27.0	14 Si 28.1	15 P 31.0	16 S 32.1	17 Cl 35.5	18 Ar 39.9
19 K 39.1	20 Ca 40.1	21 Sc 45.0	22 Ti 47.9	23 V 50.9	24 Cr 52.0	25 Mn 54.9	26 Fe 55.8	27 Co 58.9	28 Ni 58.7	29 Cu 63.5	30 Zn 65.4	31 Ga 69.7	32 Ge 72.6	33 As 74.9	34 Se 79.0	35 Br 79.9	36 Kr 83.8
37 Rb 85.5	38 Sr 87.6	39 Y 88.9	40 Zr 91.2	41 Nb 92.9	42 Mo 95.9	43 Tc (97)	44 Ru 101.1	45 Rh 102.9	46 Pd 106.4	47 Ag 107.9	48 Cd 112.4	49 In 114.8	50 Sn 118.7	51 Sb 121.8	52 Te 127.6	53 I 126.9	54 Xe 131.3
55 Cs 132.9	56 Ba 137.3	see below 57-71	72 Hf 178.5	73 Ta 180.9	74 W 183.9	75 Re 186.2	76 Os 190.2	77 Ir 192.2	78 Pt 195.1	79 Au 197.0	80 Hg 200.6	81 Tl 204.4	82 Pb 207.2	83 Bi 209.0	84 Po 210	85 At (210)	86 Rn (222)
87 Fr (223)	88 Ra (226)	see below 89-103															

57 La 138.9	58 Ce 140.1	59 Pr 140.9	60 Nd 144.2	61 Pm (145)	62 Sm 150.4	63 Eu 152.0	64 Gd 157.3	65 Tb 158.9	66 Dy 162.5	67 Ho 164.9	68 Er 167.3	69 Tm 168.9	70 Yb 173.0	71 Lu 175.0
89 Ac (227)	90 Th 232.0	91 Pa (231)	92 U 238.0	93 Np (237)	94 Pu (244)	95 Am (243)	96 Cm (247)	97 Bk (247)	98 Cf (251)	99 Es (254)	100 Fm (257)	101 Md (256)	102 No (254)	103 Lw (257)

Table 5.2: Periodic Table of the Elements

13. There is a tendency for an atom to want to have its outer electron level full. This tendency can be used to explain some of the chemical behavior of atoms. Let's look at an example.

(a)   Which atom has only two electrons? _____

(b)   Which atom has two electrons in its first orbit and eight in its next orbit (and no

other electrons)? _____

(c)   Which atom has its first three orbits full (18 electrons in all)? _____

(d)   Now refer to the periodic table. Which column are these in? _____

— — — — — — — — — — — — — — —

(a) helium; (b) neon; (c) argon; (d) far right—the noble gases

14. These gases, as you have seen, all have their outer orbits of electrons full. They are "satisfied" and have no need to share electrons with nearby atoms. They therefore do not react chemically with other atoms. The atoms of all of the elements in the column just to the left of the noble gases (7A) lack one electron of having a full outer orbit. Because of this they tend to combine with atoms having an "extra" electron in their outer orbit. It is found that atoms in the first column (1A)—lithium, sodium, potassium (K), and so on, unite with the column 7A elements in a one-to-one ratio to form molecules such as NaCl, KCl, and LiF.

With elements of which column would you expect column 2A elements to unite in

a one-to-one ratio? _____

— — — — — — — — — — — — — — —

6A (Column 2A elements have two extra electrons in their outer shells. The elements of column 6A lack two electrons from having full shells.)

ATOMIC MASSES

15. The number shown in the bottom of each box in the periodic table is the relative mass of an atom of that element. It is called the atomic mass (or atomic weight).* The unit upon which it is based is the *atomic mass unit*, or amu.

(a)   What is the average mass of a hydrogen atom? _____ amu

(b)   What is the average mass of an oxygen atom? _____

(c)   Is the atomic mass higher or lower than the atomic number? _____

— — — — — — — — — — — — — — —

(a) 1.008; (b) 16.0 amu; (c) higher

_____

*For an explanation of why most of the atomic masses listed in the table are very nearly whole numbers, as well as the reason that I refer to the "average" mass of an atom, turn to Chapter 25.

AVOGADRO'S NUMBER

16.   The atomic mass of hydrogen is about 1. The atomic mass of carbon is 12. Thus one atom of carbon has a mass 12 times as great as one atom of hydrogen. And 359 atoms of carbon have a total mass 12 times the mass of that many hydrogen atoms. And $6.023 \cdot 10^{26}$ atoms of carbon have a total mass 12 times the mass of *that* many hydrogen atoms.* This last number, $6.023 \cdot 10^{26}$ is a special one, because that many hydrogen atoms have a mass of one kilogram. It follows that $6.023 \cdot 10^{26}$ carbon atoms has a mass of 12 kilograms, then. And since the atomic mass of oxygen is 16, $6.023 \cdot 10^{26}$ oxygen atoms have a mass of 16 kilograms. This number of atoms is called Avogadro's number, and the mass of an element having that many atoms is called a *kilogram atomic mass* (or *kilogram atomic weight*) of the substance. Look up the atomic mass of chlorine in the periodic table.

(a)   One kilogram atomic mass of chlorine contains how many kilograms?_____

(b)   How many atoms of chlorine are found in one kilogram atomic mass? _____

— — — — — — — — — — — — — —

(a) 35.5; (b) $6.023 \cdot 10^{26}$

17.   Now we move from atoms to molecules. The mass of one water molecule is 18 times the mass of one hydrogen atom, since it contains two hydrogen atoms and one oxygen atom—the oxygen's atomic mass being 16. The kilogram *molecular* mass of water is therefore 18 kilograms. A kilogram molecular mass is more commonly called a *mole*.

(a)   Use the periodic chart to calculate the mass of a mole of sulfuric acid (H_2SO_4).

_____

(b)   How many molecules will this mole contain? _____

— — — — — — — — — — — — — —

(a) 98.1 amu; (b) $6.023 \cdot 10^{26}$

18.   Now try a few problems.

(a)   How many molecules are in 44 kilograms of CO_2 (carbon dioxide)?_____

(b)   How many molecules are in 11 kg of CO_2? _____

(c)   What is the mass (in kilograms) of one molecule of CO_2? (Hint: There are 44 kg in $6.023 \cdot 10^{26}$ molecules.) _____

— — — — — — — — — — — — — —

(a) $6.023 \cdot 10^{26}$
(b) $1.506 \cdot 10^{26}$ (You have $\frac{1}{4}$ mole, since $\frac{11}{44} = \frac{1}{4}$
(c) $7.305 \cdot 10^{-26}$ kilograms. Solution:

$$\frac{44 \text{ kg}}{6.023 \cdot 10^{26} \text{ molecules}} = 7.305 \cdot 10^{-26} \text{ kg/molecule}$$

*See Appendix I for an explanation of powers-of-ten notation.

SELF-TEST

The questions below will test your understanding of this chapter. Use a separate sheet of paper for your diagrams or calculations. Compare your answers with the answers provided following the test.

Refer back to the table of elements and the periodic table at any time.

1.   Identify the substances below as elements or compounds.

(a) berkelium _____ (b) ammonia _____ (c) MgO _____

(d) tin _____ (e) Hf _____

2.   What do we call the smallest unit an element can be broken into and still have the same properties? _____

3.   What do we call the smallest unit of a compound which retains the properties of the compound? _____

4.   In the figure below, mark the location of the electron, the proton, and the neutron. Mark the part of the atom which is positively charged and the part which is negatively charged.

5.   Use the periodic table to find the element with atomic number 27. _____ With atomic mass 238. _____

6.   What does the atomic number tell us about an atom? _____

7.   Explain, on the basis of electron orbits, why gases such as helium do not enter into chemical reactions. _____

_____

8.   The magnesium atom has two electrons in its outer orbit. Name two other elements whose atoms have this characteristic. _____

_____

9.   What is meant by the atomic mass of an element? _____

_____

10.   Use the periodic table to find the atomic number and atomic mass of sodium (Na).

_____

11.   How many molecules are contained in 2.8 kilograms of carbon monoxide (CO)?

_____

12.   What mass of gold (Au) is required to have 10^{25} gold atoms? _____

Answers to Self-Test

If your answers do not agree with those given below, review the frames indicated in parentheses before you go on to the next chapter.

1.   elements: a, d, and e; compounds: b and c   (frame 2)

2.   atom   (frame 4)

3.   molecule   (frame 5)

4.

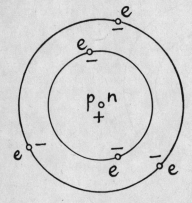

p = protons
n = neutrons
e = electrons

(frame 8)

5.   Co (cobalt); U (uranium)   (frame 11)

6.   the number of protons   (frame 10)

7.   Their outermost electron orbit is full.   (frames 13–14)

8.   Be (beryllium), Ca (calcium), Sr (strontium), Ba (barium), Ra (radium)
(frame 11)

9.   the relative mass of the atom (in atomic mass units)   (frame 15)

10.   11, 23   (frames 10, 15)

11. $6.023 \cdot 10^{25}$ molecules (From the periodic table we learn that carbon monoxide
has a molecular mass of 28, or 12 + 16. Thus we have 0.1 mole of the compound.
0.1 times Avogadro's number yields the answer.)   (frame 17)

12.   2.9 kilograms. Solution:

$$10^{25} \text{ molecules is } \frac{10^{25}}{6.023 \cdot 10^{26}} \text{ or } .0164 \text{ mole}$$

One mole of gold (atomic mass 197) has a mass of 179 kilograms, so we need .0164 times this, or 2.9 kilograms.   (frame 17)

CHAPTER SIX
Solids

Prerequisite: Chapter 5

We have daily contact with matter in three states: solid, liquid, and gas. This chapter is concerned with solids, while the next deals with liquids and gases. The solids we encounter are as different as apples and suede hats, as different as ice cubes and horseshoes. But there are similarities, too, and science progresses largely by looking for similarities among objects.

After a study of this chapter, you should be able to:

- describe the arrangement of atoms or molecules in a solid;

- differentiate between mass desntiy and weight density;

- calculate the density of an object, its weight (or mass), or its volume, given the other two quantities;

- given the specific gravity of a material, calculate its density;

- calculate the pressure exerted, given the force and area;

- describe a perfectly elastic substance;

- use Hooke's Law to determine force or deformation;

- state the conditions under which Hooke's Law applies.

ATOMS IN A SOLID

1.   When matter is in a solid state, its atoms or molecules are arranged in definite, fixed patterns.* The atoms are held in their positions by electrical forces between each atom and its neighbors. These forces give the solid its rigidity. The result of the regular arrangement of molecules is evident in the shape of snowflakes, which—although they are very

---

*In another type solid, not even called a solid in some books, the atoms and molecules have a random arrangement. Glass is of this type, called an *amorphous* solid or a *supercooled* liquid. Most of us would rather not call glass a liquid, but what about margarine—is it solid or liquid? (It is amorphous.) In this chapter, we will assume "solid" actually means "crystalline solid." (The word "crystalline" tells us definitely that the solid has a regular molecular arrangement.)

different in detail—all have a six-sided symmetry. Other arrangements of molecules give characteristic shapes for other type crystals.

(a)   Every grain of table salt has a characteristic cubic shape. What property of solids causes this? _____

(b)   What could you predict about the arrangement of molecules in something that is *not* a solid? _____

‒ ‒ ‒ ‒ ‒ ‒ ‒ ‒ ‒ ‒ ‒ ‒ ‒ ‒ ‒ ‒

(a) fixed pattern of molecular arrangement; (b) random or irregular

DENSITY

2.   Solids differ in many ways, including color, texture, and elasticity. Another important distinguishing property of solids is density. Density is defined in two different ways and to avoid confusion we will call one—the more fundamental one—"mass density," and the other "weight density."

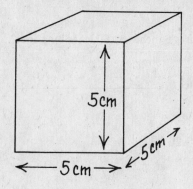

Mass density is defined as the mass of a substance divided by its volume. For example, suppose we have a steel cube 5 cm by 5 cm by 5 cm. Its mass (measured on a balance) is 975 grams.

(a)   What is its volume? _____

(b)   What is the mass density of the steel?

_____ gm/cm^3

‒ ‒ ‒ ‒ ‒ ‒ ‒ ‒ ‒ ‒ ‒ ‒ ‒ ‒ ‒

(a) 125 cm^3; (b) 7.8 (975 gm/125 cm^3)

3.   A larger piece of steel on which you make the same measurements and do the same calculations will have the same value for its mass density, since density is characteristic of the material and does not depend upon how much material is present.
    The density of aluminum is 2.7 gm/cm^3. Suppose you have 12 cm^3 of aluminum.

(a)   What is its mass? _____

(b)   What is the mass density of a piece of aluminum 24 cm^3 in volume? _____

‒ ‒ ‒ ‒ ‒ ‒ ‒ ‒ ‒ ‒ ‒ ‒ ‒ ‒ ‒

(a)   32.4 gm. Solution: density = mass/volume
                          mass = density · volume
                               = 2.7 gm/cm^3 · 12 cm^3
                               = 32.4 gm
(b)   2.7 gm/cm^3

4.   If mass is measured in kilograms and volume in cubic meters, density has units of kg/m^3. One gm/cm^3 is equivalent to 1000 kg/m^3. Table 6.1 shows the densities of some common substances.

Substance	Mass density (gm/cm^3)	Weight density (lb/ft^3)
steel	7.8	488
aluminum	2.7	168
bone	1.8–2.0	112–125
mercury	13.6	849
ice	0.92	57.4
wood	.35–.85	22–53
lead	11.3	705
water	1.00	62.4

Table 6.1

Weight density is defined as the weight of a substance divided by its volume and is normally used in the British system of units. The unit of weight density in this system is lb/ft^3 (or $lb/in.^3$).

(a)   Steel has a weight density of 488 lb/ft^3. How much does a cubic foot of steel weigh? _____

(b)   What is the weight of a block of lead which measures 6 inches by 6 inches by 1 foot? (See Table 6.1.) _____

(c)   What is the volume (in m^3) of 5000 kilograms of ice? _____

– – – – – – – – – – – – – – – –

(a)   488 lb

(b)   176 lb. Solution: volume = .5 ft · .5 ft · 1 ft = .25 ft^3
$$weight = density \cdot volume$$
$$= 705 \ lb/ft^3 \cdot .25 \ ft^3$$
$$= 176.25 \ lb$$

(c)   5.43 m^3. Solution: First, find the density in units of kg/m^3:

$$.92 \ gm/cm^3 \cdot \frac{1000 \ kg/m^3}{1 \ gm/cm^3} = 920 \ kg/m^3$$

density = mass/volume
volume = mass/density
$$= 5000 \ kg \div 920 \ kg/m^3$$
$$= 5.43 \ m^3$$

SPECIFIC GRAVITY

5.   The *specific gravity* of a substance is often listed in tables instead of either mass density or weight density. Specific gravity is defined as the ratio of the density of the substance in question to the density of water. The mass density of water* is 1.00 gm/cm^3. Thus the specific gravity of lead (density 11.3 gm/cm^3) is 11.3. (11.3 $gm/cm^3 \div 1 \ gm/cm^3$)

––––––––––––

*Notice that density is defined for liquids as well as solids—in fact, water is the reference point for specific gravity.

The specific gravity of a substance is always numerically equal to its mass density in gm/cm^3. Notice, however, that specific gravity has no units since it is simply a ratio of two densities.

The specific gravity of concrete is about 2.7. (Since concrete is a mixture that can be mixed in different proportions, this is approximate.) What is the weight of a cubic yard of concrete? (You can either work this out yourself or go through the steps below.)

(a)   What is the density of concrete in lb/ft^3? (Use Table 6.1 to find the density of

water.) _____

(b)   How many cubic feet are in a cubic yard? _____

(c)   What is the weight of a cubic yard of concrete? _____

— — — — — — — — — — — — — — —

(a)   168 lb/ft^3 (Since the weight density of water is 62.4 lb/ft^3, the density of concrete
        is 2.7 · 62.4 lb/ft^3.)
(b)   27 ft^3 (3 ft · 3 ft · 3 ft)
(c)   4540 lb (168 lb/ft^3 · 27 ft^3)
        (When you realize that this is about $2\frac{1}{4}$ tons and that concrete trucks carry from 5
        to 10 cubic yards of concrete, you realize why the trucks are so sturdily built.)

PRESSURE

6.   The total amount of force on an object is not always the critical factor. For example, you might stack two or three books on your head without feeling pain, but if you put a tiny pebble between the bottom book and your skull, the sensation could be very unpleasant. The important factor here is force divided by the area over which the force is applied, or the *pressure.*

Let's calculate the pressure exerted on the floor by a 200 lb man standing on both feet. If the contact area of each shoe is 3 by 10 inches, the area of contact is 30 in.^{2} for each shoe. What is the pressure exerted on the floor? (Include units in your answer.)

_____

— — — — — — — — — — — — — —

3.33 lb/in.^{2} (200 lb/60 in.^{2})

7.   A few more calculations of pressure will demonstrate how it can be increased.

(a)   Suppose the same man sits on a kitchen chair. Each leg of the chair might have
        about $\frac{1}{2}$ inch by $\frac{1}{2}$ inch of contact with the floor, making a total area of 1 in.^{2} for
        the four legs. What is the pressure on the floor? (Ignore the chair's weight.)

_____

(b)   In the 1950s, when women wore spike heels with areas of about $\frac{1}{16}$ in.^{2} ($\frac{1}{4}$ in. by
        $\frac{1}{4}$ in.), aircraft companies were concerned because such heels could exert enough
        pressure to punch through the floor of planes. What pressure would a 110 lb woman

        exert on the floor when she puts her entire weight on one heel? _____

— — — — — — — — — — — — — —

(a)   200 lb/in.^{2} (No wonder they damage tile floors.)
(b)   1760 lb/in.^{2} (110 lb ÷ $\frac{1}{16}$ in.^{2})

ELASTICITY

8.   Many solids can be deformed but will return to their original shape when the cause of the deformation is removed. The elasticity of a substance is a measure of how well it returns to its original shape. According to this definition we find that steel must be classified as more elastic than rubber. Although steel is not easily deformed, it returns to its original shape more readily than does rubber.

(a)   Would you classify such materials as dough and clay as elastic or inelastic? Why?

_____

(b)   Which type of substance returns exactly to its original form, perfectly elastic or perfectly inelastic? _____

_ _ _ _ _ _ _ _ _ _ _ _ _ _ _

(a)   inelastic, because they don't return to original shape
(b)   perfectly elastic (Everyday substances lie somewhere between these two extremes, but it is instructive to consider the case of perfect elasticity.)

9.   Hooke's law states that the amount of deformation produced by a force is proportional to the amount of force. Only a perfectly elastic object obeys Hooke's law perfectly. A steel spring is close enough to perfectly elastic that we can consider it so, as long as it is not stretched too far. Suppose you put a 1 pound weight on such a spring and it stretches 0.75 inches. A 2 pound weight will cause a deformation of 1.5 inches. In equation form we can write:

$$F = k \cdot x$$

where F is the force applied, k is a constant (called the spring constant), and x is the deformation. Since a force of 1 lb produces a 0.75 inch deformation, the spring constant k for the spring above is 1.33 lb/inch.

(a)   How much deformation will a 3 pound force cause? _____

(b)   How much force is required to stretch the spring 3 inches?_____

_ _ _ _ _ _ _ _ _ _ _ _ _ _ _

(a) 2.25 inches; (b) 4 pounds

10.   The spring constant tells us how "stiff" a spring is. Different springs have different spring constants.

(a)   How much deformation is produced by a force of 2 pounds if the spring constant

is 2 lb/in.? _____

(b)   Will a spring with a constant of 5 lb/in. require more or less force to stretch it a given distance than a spring with a constant of 8 lb/in.? _____

– – – – – – – – – – – – – – – –

(a) 1 inch; (b) less

SELF-TEST

The questions below will test your understanding of this chapter. Use a separate sheet of paper for your diagrams or calculations. Compare your answers with the answers provided following the test. Refer to the table of densities (page 70) at any time.

1.   Crystals of common table salt tend to have 90° angles between their flat sides. What causes this? _____

2.   A certain block of wood has dimensions of 2 in. by 4 in. by 20 in. If it weights 3.0 pounds, what is its density in lb/in.^{3}? _____

3.   There are 1728 in.^{3} in a ft^3. What is the specific gravity of the wood in the last problem? _____

4.   How much would 5 cubic feet of ice weigh? _____

5.   Why is the specific gravity of a substance equal to its density in grams/cm^3? _____

6.   Suppose the entire 3 pound weight of some books rests on your head with a wooden cube between. If the cube is $\frac{1}{4}$ inch on a side, what pressure is exerted on your head? _____

7.   Does Hooke's law apply to all objects? Why or why not? _____ _____

8.   If a spring has a spring constant of 3 pounds/inch, how far would it stretch if 10 pounds were hung from its end?_____

9.   Suppose you have 2 cubic feet of each of the materials in Table 6.1. Which will weigh the least? The most? _____

10.   A gallon of water weighs 8 lb. What is its volume in cubic feet? _____

Answers to Self-Test

If your answers do not agree with those given below, review the frames indicated in parentheses before you go on to the next chapter.

1. The regular arrangement of atoms within the crystal causes the regularity of shape. (frame 1)

2. 0.01875 lb/in.^3 Solution:

    volume $= 2 \text{ in.} \cdot 4 \text{ in.} \cdot 20 \text{ in.} = 160 \text{ in.}^3$
    density $= 3 \text{ lb} \div 160 \text{ in.}^3 = 0.01875 \text{ lb/in.}^3$ (frame 4)

3. 0.519. Solution: First find the density in lb/ft^3:

    $$0.01875 \text{ lb/in.}^3 \cdot \frac{1728 \text{ in.}^3}{1 \text{ ft}^3} = 32.4 \text{ lb/ft}^3$$

    $$\text{specific gravity} = \frac{\text{density of material}}{\text{density of water}}$$

    $$= \frac{32.4 \text{ lb/ft}^3}{62.4 \text{ lb/ft}^3}$$

    $$= .519 \quad \text{(frame 5)}$$

4. 287 lb. Solution: density $= 57.4 \text{ lb/ft}^3$
    weight $=$ density $\cdot$ volume
    $= 57.4 \text{ lb/ft}^3 \cdot 5 \text{ ft}^3$
    $= 287 \text{ lb}$ (frame 4)

5. Specific gravity is defined as the ratio of the density of the substance to the density of water. Since the density of water is 1.0 gm/cm^3, the specific gravity is numerically equal to the density.   (frame 5)

6. 48 lb/in.^2 (The area of one face of the cube is $\frac{1}{4}$ in. $\cdot$ $\frac{1}{4}$ in. or $\frac{1}{16}$ in.^{2}. The pressure is thus $3 \text{ lb} \div \frac{1}{16} \text{ in.}^2$.)   (frames 6–7)

7. No. It applies exactly only to materials which are perfectly elastic. Some materials are very inelastic, and Hooke's law does not apply to them at all.   (frame 9)

8. 3.33 inches (It requires 3 pounds to stretch an inch. Ten pounds should stretch it 10/3 times that far.)   (frame 9)

9. wood, mercury   (frame 4)

10. 0.128 ft^3 ($8 \text{ lb} \div 62.4 \text{ lb/ft}^3$)   (frames 2–4)

CHAPTER SEVEN
Liquids and Gases

Prerequisites: Chapters 5 and 6

Of the three states of matter—solid, liquid, and gas—both liquids and gases are considered fluids. Although in everyday language we often consider the word fluid to apply only to liquids, in physics gases are considered fluids also, since gases and liquids behave similarly in many situations.

After studying this chapter you will be able to:

- compare liquids and gases to solids in terms of molecular attraction and organization;

- compare liquids to solids in terms of elasticity;

- calculate pressure at a given depth in a given liquid;

- use Archimedes' principle to find the buoyant force on a given object in a given fluid;

- calculate the depth to which a given floating object will sink in a given fluid;

- use Pascal's principle to demonstrate how pressure is transmitted in a liquid;

- compare diffusion rates in liquids and gases and use kinetic theory to explain the differences;

- calculate the force on a given object due to atmospheric pressure;

- relate atmospheric pressure to "inches of mercury";

- relate the pressure of a confined gas to the number of molecules and the average speed of the molecules.

MOLECULES IN A LIQUID

1.   A liquid is distinguished from a solid in that a liquid flows to fill the bottom of a container until the liquid has a level surface. In a solid, molecules are held to a definite position in a regular pattern (although they may vibrate about a central position), while

in a liquid the molecules are free to move about one another. A force is still exerted between neighboring molecules of a liquid, for the molecules do not easily leave the surface —a liquid does not expand to fill the entire container which holds it. Nor does it contract easily; if enough pressure is exerted on a liquid to cause it to contract, its original volume is restored when the force is removed.

(a)   How are molecules arranged in a liquid—regularly or randomly? _____

(b)   Are molecules in a liquid held more or less tightly together than in a solid?

_____

(c)   Are liquids in general elastic or inelastic? _____

— — — — — — — — — — — — — — — —

(a) randomly; (b) less tightly; (c) elastic (see Chapter 6, frame 8)

PRESSURE IN A LIQUID

2.   Most of us experience the feeling of increased pressure on our ears when we dive deep below the surface of a swimming pool. Pressure in a liquid increases with depth below the surface, and can be calculated by the simple formula:

pressure = depth · weight density

(a)   What is the water pressure at a depth of 10 feet below the surface of water? (The density of water is 62.4 lb/ft^3.) _____

(b)   What is this pressure in units of $lb/in.^2$? _____

— — — — — — — — — — — — — — — —

(a) 624 lb/ft^2 (10 ft · 62.4 lb/ft^3); (b) 4.3 $lb/in.^2$ (There are 144 square inches in a square foot. So 625 lb/ft^2 is equal to 624/144, or 4.3 $lb/in.^2$.)

3.   Now assume the area of the ear drum to be $\frac{1}{16}$ $in.^2$ (which is $\frac{1}{4}$ in. by $\frac{1}{4}$ in.), and calculate the force exerted on it by the water when you are 10 feet under.

_____

— — — — — — — — — — — — — — — —

.27 lb (4.3 $lb/in.^2$ · $\frac{1}{16}$ $in.^2$)
(This small force causes pain because the ear is sensitive to very small forces.)

4.   In SI units, density is expressed as kg/m^3. This is a mass density, so we cannot simply multiply density times depth to obtain pressure. First we must calculate the weight density. Recall that one kilogram has a weight of 9.8 newtons,* and that in general we multiply mass in kilograms by 9.8 to obtain weight in newtons. Therefore, the weight density of water (1000 kg/m^3) is 9800 N/m^3. Now we'll calculate the pressure at a depth of 5 meters in alcohol (which has a specific gravity of 0.80).

_____

*If you have not studied Chapter 1, you need only know that the newton (N) is the unit of force in SI—International System—units. See frames 13, 14, and 16 of Chapter 1 for more details.

(a)   What is the mass density of alcohol? _____

(b)   What is the weight density of alcohol? _____

(c)   What is the pressure 5 meters deep in alcohol? _____

(d)   One square meter is equal to 10,000 square centimeters. What is the pressure in

 N/cm^2? _____

_ _ _ _ _ _ _ _ _ _ _ _ _ _ _ _

(a) 800 kg/m^3 (see Chapter 6, frame 5); (b) 7840 N/m^3; (c) 39200 N/m^2; (d) 3.92 N/cm^2

BUOYANCY

5.   Consider a cubic foot of water in a swimming pool full of water. (To do this you might think of the cubic foot being enclosed by thin tissue paper.) The water weighs 62.4 pounds. Since it is held there in the water of the pool, the water around it must be exerting an *upward* force of 62.4 pounds on it. If somehow the cubic foot of water were removed and an equal volume of something else put in its place, that something else would also experience an upward force of 62.4 pounds. We call this upward force the *buoyant force,* and the example above illustrates the general rule for determining the magnitude of the buoyant force: *An object in a fluid*—we will see that this principle also applies to a gas—*is buoyed up with a force equal to the weight of the fluid displaced.* This principle, called *Archimedes' principle,* is important enough to deserve some examples.

   Suppose a cubic foot of steel (weight: 480 lb) is below the surface of water.

(a)   What is the buoyant force exerted on it by the water? _____

(b)   What, then, will be the apparent weight of the steel? (In other words, how much

 force would you have to exert to lift the steel under water?) _____

_ _ _ _ _ _ _ _ _ _ _ _ _ _ _ _

(a) 62.4 lb; (b) 418 lb (480 lb minus 62 lb)
Note that although presure is greater at a greater depth, the buoyant force does not depend upon the depth.

6.   A cubic foot of pine wood weighs about 30 pounds. If it were pushed under water, the buoyant force on it would also be 62.4 lb. But since the buoyant force is greater than its weight, pine will not stay under water unless held there. Let's calculate how much of its volume will remain under water when it floats. To do this we reason that it will sink in the water until it is buoyed up by a force which is equal to its weight.

(a)   What weight of water will the 30 lb piece of wood have to displace in order to

 float? _____

(b)   What is the volume of this weight of water? _____

(c)   How much of the wood will be below water level as the block of wood floats?

 _____

_ _ _ _ _ _ _ _ _ _ _ _ _ _ _ _

(a) 30 lb; (b) 0.48 ft^3 (30 lb ÷ 62.4 lb/ft^3); (c) 0.48 ft^3

7.   In general, one can determine what fraction of a floating object will be under a liquid by dividing the density of the object by the density of the liquid.

(a)   What is the density of the wood in the last frame? _____

(b)   What is the density divided by the density of water? _____

(c)   Suppose a piece of oak wood floats with 62 percent of its volume under water.

       What is the density of the wood? _____

– – – – – – – – – – – – – – – –

(a)   30 lb/ft^3 (The cubic foot weighed 30 lb.)
(b)   .48 (Since the wood had a volume of 1 ft^3, this corresponds to answer b of the last frame.)
(c)   38.7 lb/ft^3 (or 620 kg/m^3) (This is .62 times the density of water.)

Optional

8.   As in calculating pressure, we must again use the weight density rather than the mass density in calculations involving buoyancy. Let us determine the apparent weight in water of a 5 cm^3 piece of aluminum—about pencil size. (The specific gravity of aluminum is 2.7.)

(a)   What is the weight density of aluminum in N/m^3? _____

(b)   What is the weight density in N/cm^3? _____

(c)   How much does the piece of aluminum weigh in air? _____

(d)   What is the weight density of water in SI units? _____

(e)   Convert this to N/cm^3. _____

(f)   What is the buoyant force on 5 cm^3 of aluminum? _____

(g)   What is the apparent weight of the aluminum in water? _____

– – – – – – – – – – – – – – – –

(a)   26,500 N/m^3 (2,700 kg/m^3 · 9.8 N/kg)
(b)   0.0265 N/cm^3 (There are 10^6 cm^3 in a m^3.)
(c)   .133 N (0.0265 N/cm^3 · 5 cm^3)
(d)   9800 N/m^3
(e)   0.0098 N/cm^3
(f)   0.049 N (the weight of 5 cm^3 of water)
(g)   0.084 N (0.133 N − 0.049 N)

PASCAL'S PRINCIPLE

9.   When additional pressure is put on a confined liquid, the pressure is transmitted equally to all parts of the liquid. This principle—*Pascal's principle*—is best understood by looking at an example. The figure on the next page illustrates a water container with two

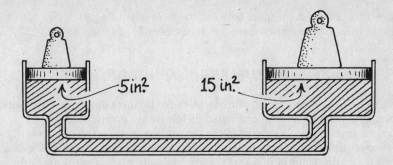

openings, each sealed by a movable piston. The openings, and therefore the pistons, have different areas, however—the left one being 5 square inches and the right one 15 square inches. The liquid just below the pistons is at the same height, so the pressure is the same below each. (Remember that pressure depends upon depth and liquid density.)

Now suppose we put a 20 pound weight on the left piston. What weight must be put on the other piston to balance this weight? To solve this, we must calculate the extra pressure exerted on the liquid due to the weight at left. This is (20 lb ÷ 5 in.^{2}), or 4 lb/in.^{2}. Now according to Pascal's principle, this pressure is exerted at all parts of the liquid, including on the piston at right. But the area of this piston is 15 in.^{2}, and so the force exerted on it by the extra water pressure is 4 lb/in.^{2} · 15 in.^{2}, or 60 lb. Thus a 60 lb weight would be required on the right to balance the 20 lb weight on the left.

The hydraulic jack uses two pistons connected basically as shown in the figure. It can be used to lift very heavy objects because its pistons have greatly different areas.

(a)   Suppose we placed a 50 lb weight on the left piston. What weight would be needed on the right to balance it? _____

(b)   In using a hydraulic jack, on which piston (smaller or larger) would a person exert his or her force? _____

– – – – – – – – – – – – – –

(a)   150 lb (10 lb/in.^{2} · 15 in.^{2})
(b)   smaller

MOLECULES IN GASES

10.   Most of us are familiar with the fact that when a liquid changes to a gas—as when water is boiled—the gas takes up much more room than the liquid. This happens because the molecules of the gas are much farther apart than the molecules of the liquid. In fact, the molecules of the gas are not bound together by forces at all, but fly about freely in space. They exert forces on each other only when they collide. When two molecules do collide they bounce from each other and continue freely until the next collision.

This model of a gas, called the kinetic theory, is consistent with a number of every-day observations concerning gases. For example:

(a)   Which can be compressed more easily, liquids or gases? _____

(b)   Which do you expect to be less dense, liquids or gases? _____

– – – – – – – – – – – – – –

(a)   gases (because there is space between the molecules of a gas)
(b)   gases (because the empty space contributes nothing to the weight)

DIFFUSION THROUGH LIQUIDS AND GASES

11.   Diffusion is the intermixing of molecules of substances due to random motion of the molecules. The molecules of one liquid diffuse rather slowly through another liquid. Suppose you carefully put a drop of ink in the bottom of a glass of water which has been sitting long enough that the water is not swirling about. (You could do this with an eye-dropper.) If you do, you will find that the ink gradually spreads out through the water. It will be hours—or even days—before the ink is entirely diffused through all of the water, however.
    Now we'll consider one gas diffusing through another. If you spill some ammonia at one end of a room, before long you can smell the ammonia halfway across the room, and a little later the smell will have filled the entire room. This smell is simply the result of molecules of ammonia entering your nose. In a matter of minutes the ammonia molecules move as far as 10 or 20 feet.

(a)   Which occurs faster, diffusion of a liquid through a liquid or a gas through a gas?

_____

(b)   What actually happens in diffusion? _____

_____

(c)   How does the molecular spacing in a gas contribute to diffusion through a gas being

faster than through a liquid? _____

_____

– – – – – – – – – – – – – – – –

(a)   gas through a gas
(b)   The molecules of one substance intermix with those of another.
(c)   There is space between gas molecules for other molecules to move through. This is not so for a liquid; in a liquid the molecules must jostle around one another.

PRESSURE IN GASES

12.   It was stated earlier that gases are more easily compressed than liquids. But if there is so much empty space between molecules, why is any effort at all required to compress them? Why do they exert pressure? Let's consider blind bumblebees flying around in a boxcar. Such bees (if they were too ignorant to land, anyway) would constantly be bumping into the walls of the car. Each collision with a wall would result in a very small push against the wall, but if enough bumblebees were present, we can imagine not only that the overall force against the wall would be significant, but that the collisions would be so frequent that the individual bumps could not be distinguished. The effect would be that of a constant force. This is what happens in a gas. Molecules, of course, are both unimaginably small and unimaginably numerous, and their constant collisions with the walls of their container produce a constant force on walls or boundaries.
    Just as with liquids, it is usually more convenient to speak of the pressure (force/area) exerted by a gas rather than the force exerted.

(a)   Suppose the number of gas molecules in an enclosed container is increased. Would you expect the pressure due to the gas to increase or decrease? _____

(b)   Suppose the average speed of the molecules of a confined gas is increased.* How would this affect the pressure? _____

_ _ _ _ _ _ _ _ _ _ _ _ _ _ _

(a)   increase (because more molecules would be hitting the walls)
(b)   it would increase (because each collision with a wall would exert more force on the wall, and the molecules would hit the walls more frequently)

13.   The gas pressure we are most familiar with is that of the atmosphere. Because of the tremendous amount of air above us, the atmosphere exerts a force of about 14.7 pounds on every square inch of surface at sea level. Hold your hand horizontally in fron of you. The area of your hand is about 20 square inches.

(a)   What force is the atmosphere exerting downward on your hand? _____

(b)   But the air under your hand also exerts a pressure upward. What force is exerted on the underside of your hand? _____

_ _ _ _ _ _ _ _ _ _ _ _ _ _ _

(a)   294 lb (20 in.^{2} · 14.7 lb/in.^{2})
(b)   294 lb
(The hand is not crushed by these two forces because we grow up with atmospheric pressure on us and equal pressures are exerted on our flesh from the inside by the liquids and solids within our bodies.)

14.   The pressure in a gas (as in a liquid) increases as one gets deeper and deeper in the gas. Recall that in a liquid, as one doubles his depth below the surface he doubles the pressure. The case is not quite so simple for atmospheric pressure, however. (After all, what is meant by the "surface" of the atmosphere?) The density of a liquid is the same throughout; the atmosphere, however, is less dense at greater height. Thus the weight of atmosphere above one's head does not vary in any direct way with his height above sea level. The change in air pressure as one changes height is experienced, however, when we drive quickly up or down a long hill, or better yet when we ascend or descend in an airplane. We feel the same kind of discomfort in our ears as when we dive too deeply in a swimming pool.
       The atmospheric pressure at sea level is 14.7 lb/in.^{2}.

(a)   Would you expect it to be greater or less on a Colorado mountain top? _____

(b)   Would you expect the air to be more or less dense on that mountain top?

       _____

_ _ _ _ _ _ _ _ _ _ _ _ _ _ _

(a) less; (b) less

_____

    *Raising the temperature of a gas increases the average speed of the molecules. This will be discussed in Chapter 8.

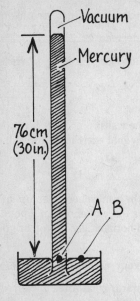

Vacuum

Mercury

76cm
(30in.)

A B

THE BAROMETER

15.   A barometer is a device used to measure atmospheric pressure. The most basic type of barometer is
shown in the figure at the left. A long tube is filled
with mercury and inverted in a pool of mercury. The
mercury does not stay in the top of the tube but
falls, leaving a vacuum at the top. The mercury level
will fall to a height of about 30 inches (about 76 cm)
above the mercury pool when the apparatus is at sea
level no matter how much longer than 30 inches the
tube is made.

The reason that the mercury stays up in the
tube is that the atmosphere, pushing down on the
surface of the mercury pool, pushes it up. And the
atmospheric pressure is able to hold the mercury up
to a height such that the pressure at point A due to
the mercury above is equal to atmospheric pressure.
(The pressure at A must equal the pressure at B because the points are at the same level.) Let's calculate
the pressure due to a height of 30 inches of mercury. The density of mercury is 0.49
$lb/in.^3$.

(a)   · What is the formula for pressure at a depth in a liquid? _____

(b)   What pressure does the barometer read at 30 inches?_____

(c)   Would a reading of 29 inches indicate more or less pressure?_____

— — — — — — — — — — — — — — — —

(a) pressure = depth · density; (b) 14.7 $lb/in.^2$ (0.49 $lb/in.^3$ · 30 in.); (c) less

16.   Rather than stating the pressure of the atmosphere in pounds per square inch, we
normally speak of the pressure being so many inches of mercury. To be absolutely correct,
we should not say that the pressure today is 29.8 inches, but that it is equal to the pressure exerted at a depth of 29.8 inches in mercury. (Weather reports on TV are limited as
to time available, so the weatherperson takes a shortcut!)

Compact barometers have an airtight container with a flexible end. As atmospheric
pressure increases, this flexible end is pushed in and a lever connected to it moves a
pointer on a dial on the face of the barometer. Because of the common use of the mercury barometer, such dials are normally calibrated in inches or centimeters of mercury
even though they contain no mercury.

(a)   What is the equivalent of 30 inches of mercury in mm Hg (millimeters of mercury)?

_____

(b)   When the pressure is 15 $lb/in.^2$, what would a mercury barometer read in inches?

_____

— — — — — — — — — — — — — —

(a)   762 mm Hg (There are 25.4 mm in an inch.)

(b)   30.6 in.   Solution: pressure = depth · density

$$\text{depth} = \text{pressure/density}$$
$$= 15 \text{ lb/in.}^2 \div .49 \text{ lb/in.}^3$$
$$= 30.6 \text{ in.}$$

BUOYANCY IN GASES

17.   Archimedes' principle states that an object in a fluid is buoyed up with a force equal to the weight of fluid displaced. We notice the effect of buoyant force in liquids more than in gases because the density of liquids is so much greater. The weight of liquid displaced is greater than the weight of gas displaced by a given object. The density of air at sea level is about 0.075 lb/ft^3, and the volume of a 125 lb person is about 2 cubic feet.

(a)   What is the buoyant force on this person due to the atmosphere? _____

(b)   What would be the buoyant force on this person in water? _____

– – – – – – – – – – – – – – –

(a)   0.15 lb or 2.4 ounces   (0.075 lb/ft^3 · 2 ft^3)

(b)   125 lb   (62.4 lb/ft^3 · 2 ft^3. The fact that this is also the weight of the person confirms the observation that people are just between floating and sinking when in water.)

18.   The most visible example of the buoyant force of air is the case of helium balloons and blimps. A large blimp may have a volume of a few million cubic feet. At 0.075 lb/ft^3, a million cubic feet displaces 75,000 pounds. In order for a million cubic foot airship to be lifted by buoyant force, all that is required is that the entire ship—structure and enclosed gas—weigh less than 75,000 pounds. The use of a very low density gas allows this to be achieved. The least dense gas is hydrogen, but it is very explosive, so helium is used. The Hindenburg was a 7 million cubic foot airship containing hydrogen; it exploded and burned in 1937.

(a)   What is the most the Hindenburg could have weighed and still flown? _____

(b)   Why do hot air balloons rise? _____

– – – – – – – – – – – – – – –

(a) 525,000 lb; (b) hot air is less dense than cold air

SOLIDS, LIQUIDS, AND GASES REVIEWED

19.   Over the last two chapters we have studied the three states of matter and have seen similarities as well as differences. Of the three states—solids, liquids, and gases—which:

(a)   are least dense (in general)? _____

(b)   have most rigid molecular bonding? _____

(c)   are most compressible? _____

(d)   permit diffusion? _____

– – – – – – – – – – – – – – –

(a) gases; (b) solids; (c) gases; (d) liquids and gases
(Solids do diffuse through solids, but centuries must pass before the effect can be seen, and even then the effect is very slight.)

SELF-TEST

The questions below will test your understanding of this chapter. Use a separate sheet of paper for your diagrams or calculations. Compare your answers with the answers provided following the test.

1.   What is the difference between a liquid and a solid as to how the molecules are

bound? _____

_____

2.   What is the relationship between pressure, depth, and density in a liquid? _____

_____

_____

3.   Calculate the pressure at a depth of 20 feet below the surface of water (density

62.4 lb/ft^3). _____

4.   A basketball has a volume of about .52 cubic feet. What is the buoyant force on it

if it is pushed under water? _____

5.   A certain cornerstone for a building is 1 ft by 1 ft by 2 ft, and weights 425 lb. What

will be its apparent weight under water? _____

6.   Will the cornerstone of question 5 apparently weigh more in a vacuum or in air?

_____

7.   In the figure, the piston at left has an area of 6 square inches and the one at right
21 square inches. What weight is needed on the left to balance 500 pounds on the

right? _____

8.   What relationship does a molecule of a gas have to its neighboring molecules?

_____

9. Suppose that two gases are in identical containers and that the gases have the same number of molecules with the same average speed. The molecules of one gas are more massive than those of the other, however. Which will exert the greater pressure on the walls of its container? Explain. _____

    _____

10. When at sea level, a pressure of about 15 lb/in.^{2} is exerted on us. How deep must we go in water so that the water will exert this much pressure? (The density of water is 0.036 lb/in.^{3}.) _____

11. Match the units (on the right) with the quantities that they are used to express.

    (a)   pressure _____   lb/ft^3

    (b)   force _____   lb/in.^{2}
                                                             N/m^2
    (c)   weight density _____   kg/m^3

    (d)   mass density _____   mm Hg
                                                             lb

12. What is the pressure (in SI units) at a depth of 3 meters in alcohol (specific gravity 0.80)? _____

13. How deep below the surface of a pool of mercury would the pressure be equal to the pressure 20 meters deep in water? (The specific gravity of mercury is 13.6.)

    _____

14. When a weather report gives the pressure as 28.9 inches, what is the pressure in lb/in.^{2}? (The density of mercury is 0.49 lb/in.^{3}.) _____

15. *Optional.* Mercury has a specific gravity of 13.6. What buoyant force is exerted on a 100 cubic centimeter object submerged in mercury? (Your answer should be in newtons.) _____

Answers to Self-Test

If your answers do not agree with those given below, review the frames indicated in parentheses before you go on to the next chapter.

1. In a solid the molecules are restricted to definite position, and cannot change places with other molecules. In a liquid the molecules exert forces on one another so that distances between molecules are pretty well fixed, but they are able to move around one another.   (frame 1)

2. pressure = depth · weight density   (frame 2)

3. 1250 lb/ft^2 (20 ft · 62.4 lb/ft^3)   (frame 2)

4. 32.4 lb. The volume of water displaced is 0.52 ft^3. The weight of this much water is 0.52 ft^3 · 62.4 lb/ft^3 = 32.4 lb.   (frame 6)

5.   300 lb.   volume of block $= 2 \text{ ft}^3$
      weight of water displaced $= 2 \text{ ft}^3 \cdot 62.4 \text{ lb/ft}^3 = 125 \text{ lb}$
      apparent weight $= 425 - 125 = 300 \text{ lb}$ (frame 5)

6.   in a vacuum (When in air, it is buoyed up slightly—0.15 lb—by the buoyant force of the air.)   (frame 17)

7.   143 lb. Solution: The pressure on the right side is $\dfrac{500 \text{ lb}}{21 \text{ in.}^2} = 23.8 \text{ lb/in.}^2$. This
pressure is exerted on the left piston, so the force exerted there is:

      force $= 23.8 \text{ lb/in.}^2 \cdot 6 \text{ in.}^2 = 143 \text{ lb}$ (frame 9)

8.   Hardly any! Molecules of a gas exert forces on one another only when collisions occur.   (frame 10)

9.   The gas with larger molecules. Pressure is due to collisions with the walls. If the molecules are moving at the same speed, the more massive molecules will exert more force in a collision.   (frame 12)

10.   420 in. or 35 ft. Solution: pressure $=$ depth $\cdot$ weight density
      depth $=$ pressure $\div$ weight density
          $= 15 \text{ lb/in.}^2 \div 0.036 \text{ lb/in.}^3$
          $= 420 \text{ inches}$ (frame 2)

11.   pressure: lb/in.^2, N/m^2, and mm Hg   (frames 15–16)
    force: lb   (frame 4)
    weight density: lb/ft^3 (frame 4)
    mass density: kg/m^3 (frame 4)

12. $22{,}500 \text{ N/m}^2$. mass density of alcohol $= .80 \cdot 1000 \text{ kg/m}^3 = 800 \text{ kg/m}^3$
      weight density of alcohol $= 800 \text{ kg/m}^3 \cdot 9.8 \text{ N/kg} = 7840 \text{ N/m}^3$
      pressure $=$ depth $\cdot$ weight density
          $= 3 \text{ m} \cdot 7840 \text{ N/m}^3 = 22{,}500 \text{ N/m}^2$ (frame 4)

13.   14.7 m. Since for a given pressure, depth and density are inversely proportional, we will work this by considering that mercury is 13.6 times more dense than water, so it will have the same pressure at 1/13.6 of the depth.

      $\dfrac{1}{13.6} \cdot 20 \text{ m} = 1.47 \text{ m}$ (frame 3)

14. 14.2 lb/in.^2.   pressure $=$ depth $\cdot$ weight density
          $= 28.9 \text{ in.} \cdot 0.49 \text{ lb/in.}^3$
          $= 14.2 \text{ lb/in.}^2$ (frame 16)

15.   13.3 newtons. Solution: The density of mercury is 13.6 times the density of water. This is 13.6 gm/cm^3. The mass of mercury displaced by the 100 cm^3 object is 1360 grams, or 1.36 kg. The weight of 1.36 kg is $1.36 \text{ kg} \cdot 9.8 \text{ N/kg}$, or 13.3 N. (frame 8)

CHAPTER EIGHT
Temperature and Heat

Prequisites: Chapters 5, 6 (frame 1), and 7 (frames 1 and 10)

Before we can define temperature or discuss how it is measured, we need to talk about three more-or-less common temperature scales: Fahrenheit, Celsius (or Centigrade), and Kelvin (or Absolute). Then after seeing how temperature is measured, we will be ready to discuss the real meaning of the term and distinguish between temperature and heat (which are often confused in everyday language).

When you finish this chapter you will be able to:

- specify reference points for the three temperature scales;

- convert readings from any of the three temperature scales to any other;

- explain the difference in operation between a mercury thermometer and one made with a bimetallic strip;

- using a table of thermal coefficients, calculate change in length due to a given change in temperature of a given solid;

- differentiate between temperature, heat, and internal energy;

- calculate the number of calories or Btu's to raise water a given number of degrees;

- use a table of specific heat to determine how much heat is needed to cause a given temperature change in a given substance.

THE FAHRENHEIT TEMPERATURE SCALE

1.   The Fahrenheit temperature scale is defined on the basis of two points: the freezing point of pure water is set at 32°F, and the boiling point of water is 212°F.

(a)   How many Fahrenheit degrees* are there between the freezing and boiling points of

pure water? _____

---
*"Degrees Fahrenheit" refers to the actual reading on the temperature scale. "Fahrenheit degrees" refers to a number of degrees change. Thus there are 5 *Fahrenheit degrees* between 40 *degrees F* and 45 *degrees F*.

(b)   What is the status of water at 0°F? _____

– – – – – – – – – – – – – – – –

(a) 180; (b) it is ice (32° below freezing)

THE CELSIUS SCALE

2.   The Celsius (or Centigrade) scale, although just as arbitrarily defined as the Fahrenheit scale, is part of the metric system. It is the only worldwide system. It sets its two fixed points at 0°C for the freezing point of water and 100°C for the boiling point.

(a)   How many Celsius degrees are there between the freezing point and boiling point of

water? _____

(b)   Compare your answer to part (a) here to that in frame 1. Which is larger—a Fahrenheit degree or a Celsius degree? _____

– – – – – – – – – – – – – – – –

(a) 100; (b) Celsius degree

3.   A Celsius degree is equal to 180/100, or 9/5, times a Fahrenheit degree. Knowing this, the only other thing that one must consider is that there are 32 Fahrenheit degrees difference in the location of the freezing point. Thus, to convert °F to °C:

$$°C = (°F - 32) \cdot \frac{5}{9}$$

Rewriting this to convert °C to °F:

$$°F = (°C \cdot \frac{9}{5}) + 32$$

(a)   What is 77°F on the Celsius scale? _____

(b)   What is 35°C on the Fahrenheit scale? _____

(c)   What is −40°C on the Fahrenheit scale? _____

– – – – – – – – – – – – – – – –

(a)   25°C: $(77 - 32) \cdot \frac{5}{9} = 45 \cdot \frac{5}{9} = 25$
(b)   95°F: $(35 \cdot \frac{9}{5}) + 32 = 63 + 32 = 95$
(c)   −40°F: $(-40 \cdot \frac{9}{5}) + 32 = -72 + 32 = -40$ (This is the only place where
       °C = °F!)

THE KELVIN SCALE

4.   It has been found that the coldest possible temperature is −273.15°C. This leads to our last scale, which has its zero point located at this coldest temperature. Then rather than define another fixed point, we simply say that one division on this scale is equal to one division on the Celsius scale. Thus the freezing point of pure water on the Kelvin scale is 273.15K. (There is no degree symbol used with this scale; this is read "273.15 Kelvins.")

       The conversion between Celsius and Kelvin is easy: one simply adds 273.15 to the Celsius temperature.

(a)   What is the boiling point of water on the Kelvin scale? _____

(b)   What is 95°F on the Kelvin scale? _____

(c)   What is 500K on the Celsius scale? _____

- - - - - - - - - - - - - - - -

(a) 373.15K; (b) 308.15°C (First, change to °C: 35.15°C.) (c) 226.85°C

5.   The Kelvin scale is not quite so arbitrary as the other two scales, because it does start at a real starting point in nature, the coldest possible temperature. It is used almost entirely, however, by the scientific community. Figure 8.1 compares the three scales.

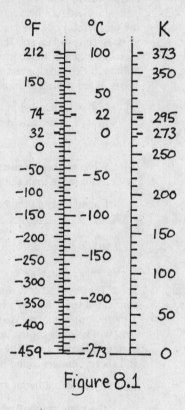

Figure 8.1

Use the figure to determine the following.

(a)   What is 98.6°F on the Celsius scale? _____

       On the Kelvin scale? _____

(b)   22°C is what temperature Fahrenheit? _____

(c)   What is the Fahrenheit reading of the coldest temperature possible? _____

- - - - - - - - - - - - - - - -

(a) 37°C, 310 K; (b) 72°F—a comfortable temperature; (c) −459°F

EXPANSION WITH TEMPERATURE: THERMOMETERS

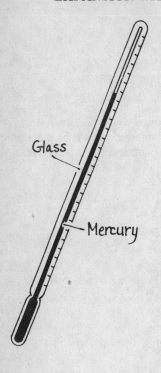

6.   Most materials expand when their temperature increases. Allowance for this expansion is made in concrete pavement by locating spaces every so often for the concrete to expand into during hot weather. (Such spaces are usually filled with a relatively soft material such as asphalt.) What is not quite so well known is that different materials expand different amounts with the same temperature rise. This fact allows the common mercury thermometer to register temperature.

In the figure here, consider what happens when the temperature of the materials of the thermometer rises. Both the glass and the mercury expand, but the mercury expands much more. If the mercury in the bulb is to expand, it must move up the thin column. Marks along the column indicate the temperature.

(a)   If glass expanded more than mercury when heated, what would happen to the level of the

mercury in the column? _____

(b)   When the temperature falls, mercury contracts, again more than the glass does. What happens to

the mercury in the column now? _____

— — — — — — — — — — — — — — — — —

(a) it would descend; (b) it falls

7.   The figure at left shows two different metals connected together to form a *bimetallic strip*. Suppose the upper metal expands more when heated. The strip will bend as shown in the lower figure. Such bimetallic strips are at the heart of oven thermometers and thermometers found in most wall thermostats.

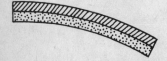

(a)   What would happen if the two metals expand

the same amount when heated? _____

_____

(b)   How would the strip bend if the lower metal expanded more when heated?

_____

— — — — — — — — — — — — — —

(a) it wouldn't bend; (b) it would curve upward—with the lower strip on the outside of the curve

8.   The amount of expansion of most materials is very nearly proportional to the rise in temperature. For example, if an iron rod expands one millimeter when its temperature rises 15°C, then it will expand 2 mm with a 30 degree rise in temperature. The equation used to calculate expansion due to temperature change is:

$$\Delta \ell = \alpha \cdot \ell \cdot \Delta T$$

where $\Delta \ell$ is the change in length and ΔT is the change in temperature.* The other symbol, α (alpha), is called the thermal coefficient of linear expansion, and is a factor which is a property of the material being considered. Table 8.1 gives the coefficients of expansion for a number of common materials.

Material	α (per $C°$)
aluminum	0.000025
brass	0.000018
glass (pyrex)	0.000003
iron	0.000011

Table 8.1

The other quantity on the right side of the equation above, ℓ, is the original length of the object, and must be taken into consideration.

Suppose we have an iron railroad rail 15 meters long. We will calculate its expansion when the temperature changes from −5°C to 25°C. (This corresponds to a change from 23°F to about 78°F—"everyday" temperatures.)

(a)   How much does the temperature change on the Celsius scale? _____

(b)   How much does the rail expand? _____

(c)   Express this expansion in millimeters. _____

– – – – – – – – – – – – – – – – –

(a)   30 Celsius degrees
(b)   0.00495 m. Solutuion: $\Delta \ell = (.000011/C°) \cdot 15 \text{ m} \cdot 30 \text{ } C°$
                                                    $\Delta \ell = 0.00495 \text{ m}$
(c)   4.95 mm

9.   Now you have a brass statue which is 3 feet high. How much will its height increase if, after being at room temperature (about 20°C), boiling water is constantly poured over it? We can mix British and metric units here because the coefficient of expansion times the temperature change is equal to the ratio of the change in length to the original length. And a ratio has no units.

Fill in the values in the equation:

$$\Delta \ell = \alpha \cdot \ell \cdot \Delta T$$

$\Delta \ell = $ _____ $\cdot$ _____ $\cdot$ _____

– – – – – – – – – – – – – – – – –

_____

*The symbol Δ, the Greek letter delta, is used to represent a change in a quantity. Thus on the left side of the equation, $\Delta \ell$ means "change in length."

$\Delta \ell = .000018/C° \cdot \underline{3 \text{ ft}} \cdot \underline{80 \text{ C}°} = .00432 \text{ ft (about .05 inch)}$

TEMPERATURE, INTERNAL ENERGY, AND HEAT

10.   Thus far we have seen how to measure temperature and how a change in temperature results in a change in the size of objects. What are we actually measuring when we measure temperature? Recall that in all materials, molecules are in constant motion. In a solid, this motion is vibration of the molecules, while in fluids the motion is that of the molecules moving from one place to another. Motion, however, implies energy—kinetic energy. Temperature is simply the measure of the average kinetic energy of the molecules. In an object with a greater temperature, the molecules have more kinetic energy and therefore greater speed.

Internal energy refers to the total kinetic and potential energies of the molecules of an object. (The potential energy results from the forces between molecules of the object.)

(a)   Which molecules have greater average kinetic energy, those in a cup of water at 40°C or in one at 20°F? _____

(b)   Which molecules have greater average kinetic energy, those in a cup of water at 40°C or those in a quart of water at 40°C? _____

(c)   What causes the potential energy of the molecules of a substance? _____

— — — — — — — — — — — — — — —

(a) 40°C; (b) the same; (c) forces between molecules

11.   Temperature involves only kinetic energy while internal energy involves both kinetic and potential energy. A second important distinction between the two is that temperature measures the *average* kinetic energy of the molecules, and internal energy refers to the *total* energies of all the molecules of an object. Suppose two cups of water, each 40°C, are poured together.

(a)   What is the temperature (due to the average kinetic energy) of the mixed water?

_____

(b)   How does the internal energy of the mixture compare to the internal energy of one

of the cups before mixing? _____

— — — — — — — — — — — — — —

(a) 40°C (the same); (b) it is twice as much

12.   Heat is a term which is often confused with temperature (and sometimes with internal energy). Heat is the internal energy which is transferred from one object to another as a result of a difference in temperature between the two objects. Note that heat is not "in" an object, but is energy that is being transferred from one object to another. Methods of heat transfer are a subject for the next chapter. For now it is sufficient that you realize that when objects of different temperature are put in contact (or mixed) with one another, they come to the same temperature by a transfer of heat from the hot to the cold object.

Identify each description below as representing heat, temperature, or internal energy:

(a)   Average kinetic energy of molecules: _____

(b)   Transfer of energy: _____

(c)   Total energy in an object: _____

- - - - - - - - - - - - - - -

(a) temperature; (b) heat; (c) internal energy

THE CALORIE

13.   Although heat and temperature are different things, they are obviously related. Usually, when we add heat to an object, its temperature rises. This leads to the definition of the calorie.* One calorie is the amount of heat which, when added to one gram of water, raises the temperature of the water one Celsius degree.

(a)   Suppose you have 15 grams of water and add 60 calories to it. How much does its

temperature rise? _____

(b)   How much heat must be added to raise 100 grams of water from 5°C to 95°C?

_____

- - - - - - - - - - - - - - -

(a) 4 Celsius degrees (60/15); (b) 9000 calories

THE BRITISH THERMAL UNIT

14.   In the British system, the unit of heat is the British thermal unit, the Btu. One Btu is the amount of heat which will raise the temperature of one pound of water one Fahrenheit degree. Note the parallel in the definitions of the calorie and the Btu.

You may hear of an air conditioner rated as, say, 12,000 Btu's. This makes no sense! Are we to suppose that the air conditioner is capable of exchanging 12,000 Btu's of heat energy before it "expires"? No. In fact, the rating is 12,000 Btu's *per hour.* The last two words are usually omitted, probably because so few people know what is meant anyway. Ask an air conditioner salesman.

(a)   How many Btu's are required to raise the temperature of 5 pounds of water 10

Fahrenheit degrees? _____

(b)   What does a rating of 12,000 Btu's per hour mean? _____

_____

- - - - - - - - - - - - - - -

(a) 50 Btu's; (b) 12,000 Btu's of energy can be exchanged each hour by the device.

---

*Since heat is energy, the calorie is a unit of energy. When this word is used to rate the energy content of food, what is really meant is the kilocalorie, which is 1000 of the calories we are defining here. It is unfortunate that we refer to the food kilocalorie as a calorie, but confusion seldom results since the meaning is usually clear from the context.

SPECIFIC HEAT

15.   When one calorie of heat is added to one gram of water, the water's temperature rises one Celsius degree. To raise one gram of glycerine one degree, on the other hand, requires only .06 calorie. And to raise one gram of aluminum one degree requires only 0.22 calorie. The ratio of "the amount of heat required to raise one gram of a substance one degree" to "the amount required to raise one gram of water one degree" is defined as the *specific heat* of the substance. Table 8.2 gives the specific heats of various substances.

Material	Specific Heat
aluminum	0.22
copper	0.09
glass	0.16
glycerine	0.60
ice	0.50
iron	0.11
silver	0.06
steam	0.50
water	1.0000

Table 8.2

Since water requires one calorie per gram degree Celsius, the specific heat of a substance is equal to the number of calories required to raise the temperature of one gram of the substance one Celsius degree. The relationship can best be expressed as a formula:

$$H = m \cdot c \cdot \Delta T$$

where H is the heat added (or removed), m is the mass of the substance, c is its specific heat, and ΔT is the change in temperature.

(a)   How much heat is required to raise the temperature of 150 grams of iron from 20°C to 25°C? _____

(b)   Which of the materials of Table 8.2 will show the greatest increase in temperature if the same amount of heat is added to one gram of each of them? _____

(c)   If 120 calories of heat are added to 100 grams of glass, at 10°C, what will be the final temperature of the glass? _____

– – – – – – – – – – – – – – –

(a)   82.5 cal. $H = m \cdot c \cdot \Delta T$
$$= 150 \cdot .11 \cdot (25 - 20)$$
$$= 82.5 \text{ cal}$$

(b)   silver (Silver requires the least amount of heat to raise its temperature a given amount. Thus if the same amount of heat is added to each material, the silver will show the greatest temperature change.)

(c)   22.5°C. $H = m \cdot c \cdot \Delta T$

$$\Delta T = \frac{H}{m \cdot c}$$

$$= \frac{120 \text{ cal}}{100 \text{ gm} \cdot .16}$$

$$= 7.5 \text{ degrees}$$

Thus, the new temperature $= 15° + 7.5° = 22.5°$

SELF-TEST

The questions below will test your understanding of this chapter. Use a separate sheet of paper for your diagrams or calculations. Compare your answers with the answers provided following the test. Refer to Figure 8.1 (page 89) at any time.

1.   State the freezing point and the boiling point of water on each of the three common temperature scales. _____

_____

2.   Change 68°F to its Celsius temperature. _____

3.   Forty degrees Celsius is how many degrees Fahrenheit? _____

4.   Forty degrees Celsius is how many Kelvins? _____

5.   Suppose a liquid-in-glass thermometer were constructed with a liquid with the same coefficient of expansion as the glass. What would happen when the temperature of the thermometer is increased? _____

_____

6.   How does a bimetallic strip respond to temperature change? _____

_____

7.   How much will a 200 ft aluminum wire expand if it is heated from 20°C to 60°C? (Use Table 8.1.) _____

8.   Will the aluminum of question 7 expand more or less than a 200 ft brass wire under the same temperature change? _____

9.   If 30 Btu's of heat are added to 5 pounds of water, how much will the temperature of the water change? _____

10.   How much heat must be added to 100 grams of copper to raise its temperature 5 Celsius degrees? (Use Table 8.2.) _____

11.   How does the heat required to raise the temperature of water a certain amount compare to the heat required to raise the same mass of ice the same amount?

Answers to Self-Test

If your answers do not agree with those given below, review the frames indicated in parentheses before you go on to the next chapter.

1. | Scale | Freezing point | Boiling point |
|---|---|---|
| Fahrenheit | 32° | 212° |
| Celsius | 0° | 100° |
| Kelvin | 273.15° | 373.15° |

(frames 1, 2, 4, and 5)

2. $20°C$: $(68 - 32) \cdot \frac{5}{9} = 20$ (frame 3)

3. $104°F$: $(40 \cdot \frac{9}{5}) + 32 = 104$ (frame 3)

4.   313K: $40 + 273.15 = 313.15$ (frame 4)

5.   The liquid would neither rise nor fall, because it would expand just as much as the glass.   (frame 6)

6.   The strip bends so that the metal which expands more is on the outside of the curve. (frame 7)

7.   0.2 ft. $\Delta \ell = \alpha \cdot \ell \cdot \Delta T$
$= .000025 \cdot 200 \cdot (60 - 20)$ (frame 8)

8.   The aluminum expands more than the brass because its coefficient of expansion is greater.   (frame 8 and Table 8.1)

9.   six Fahrenheit degrees (Five pounds of water requires 5 Btu's to raise it one degree; 30 Btu's raise it six times as far—6 degrees.   (frame 14)

10.   45 cal. $H = m \cdot c \cdot \Delta T$
$= 100 \cdot .09 \cdot 5$ (frame 15)

11.   It requires twice as much heat to raise water's temperature than it does to raise ice's (assuming the same amount of each substance to be raised by the same temperature), because the specific heat of water (1.00) is twice that of ice (0.50). (frame 15)

CHAPTER NINE

Change of State
and Transfer of Heat

Prerequisites: Chapters 5–8

In the last chapter we saw that, given the specific heat of a substance, we can calculate how much heat is required to raise the substance's temperature a certain amount. We ignored the fact that, if enough heat is added, a solid may melt or a liquid may vaporize. We will take up these phenomena in this chapter, and then will discuss how heat is transferred from one object to another.

When you complete this chapter, you should be able to:

- specify the heat of fusion of water;

- specify the heat of vaporization of water;

- calculate the amount of heat required to change ice, water, or vapor at any temperature to any other state or temperature;

- explain why amorphous solids have no heat of fusion;

- differentiate among the three ways heat is transferred;

- calculate the heat conducted through a given object, given its conductivity, its dimensions, the time elapsed, and the temperature differential;

- relate the color of an object to its ability to absorb and emit thermal radiation;

- select the predominent method of heat transfer involved in various situations.

MELTING: THE HEAT OF FUSION

1.   The freezing point of water is 0°C. At this temperature, water may be a liquid or it may be a solid—ice. The amount of internal energy determines in which of these states the water will be. If we add heat to ice which is at the freezing point (or melting point, which is the same thing), the ice will become water at that same temperature. The amount of heat required to change one gram of ice at 0°C to one gram of water at 0°C is 80 calories. This is called the *heat of fusion* of water. Other substances generally have a lower heat of fusion than water does. For example, lead's heat of fusion is 5.9 calories per gram.

(a)   How many calories are needed to change 10 grams of ice at $0°C$ to 10 grams of water at $0°C$? _____

(b)   How many calories are needed to change 10 grams of solid lead at its melting point to liquid at that temperature? _____

– – – – – – – – – – – – – – – –

(a) 800; (b) 59 (Thus lead is easier to melt—once you reach its melting point of $327.5°C$ or $621°F$.)

2.   In a crystalline solid (the type we have been discussing), the atoms are held in a definite pattern by forces exerted by neighboring molecules. Since the same pattern of molecular arrangement is repeated throughout the solid, the bonds holding one molecule in place are similar to the bonds holding others. This results in the same energy being required to break one bond as to break any other. As a solid is heated, the molecules vibrate with more and more energy (recall that temperature is the average kinetic energy of the molecules), and at some point any additional energy will begin to break their bonds. The energy that is required to break these bonds is the heat of fusion.

An amorphous solid (such as butter or glass) does not have a regular pattern of molecular arrangement, so different molecules are held with forces of different strength. Thus as the solid is heated, some bonds break at low temperatures and some at higher temperatures. This results in the gradual softening of such substances rather than a sudden melting at a given, fixed temperature (as is the case with crystalline solids).

(a)   Would you expect margarine to be a crystalline or amorphous solid? _____

(b)   Which have no heat of fusion, crystalline or amorphous solids? _____

(c)   Which have molecular bonds of varying strengths, crystalline or amorphous solids?
       _____

– – – – – – – – – – – – – – – –

(a) amorphous (since it softens gradually rather than melting at a definite temperature);
(b) amorphous; (c) amorphous

THE HEAT OF VAPORIZATION

3.   Suppose you start with water at room temperature—about $20°C$—and add heat. Each calorie of heat added will increase the temperature of a gram of water one degree. Then when the water reaches $100°C$, it will not increase further in temperature, but will start to boil. Boiling is the process by which the water changes to its gaseous state, water vapor. To change one gram of water at $100°C$ to water vapor at $100°C$ requires 540 calories of heat.

(a)   Which requires more calories per gram, melting ice, or vaporizing water? _____
       _____

(b)   How much heat is needed to change 10 grams of water at $100°C$ entirely to water vapor? _____

– – – – – – – – – – – – – – – –

(a) vaporizing water; (b) 5400 calories

4.   Let's try a more complex problem: Given that the specific heat (c) of ice is 0.5, and the specific heat of water vapor is also 0.5, how much heat is required to change one gram of ice at $-20°C$ to water vapor at $130°C$? Fill in the steps below.

(a)   The first step is to find out how much heat is needed to change the ice at $-20°C$ to ice at $0°C$: $H =$ $m \cdot c \cdot \Delta t$ What is your answer?   _____ calories

Total so far: (_____)

(b)   For the next step, change the ice to water at the same temperature.   _____ calories

Total calories so far: (_____)

(c)   Now heat the gram of water from $0°C$ to $100°C$:   _____ calories

(d)   The next step is to vaporize the water.   _____ calories

Total thus far: (_____)

(e)   Finally, use the equation in (a) to calculate the heat needed to raise the temperature of the steam from $100°C$ to $130°C$.   _____ calories

(f)   What is the total number of calories needed?   _____ calories

_ _ _ _ _ _ _ _ _ _ _ _ _ _ _ _

(a) 10 (Chapter 8, frame 15); (b) 80 (total 90); (c) 100; (d) 540 (total 730); (e) 15; (f) 745 calories

5.   In a liquid, molecules are free to move around one another, but there are forces which prohibit them from separating from one another. At any given temperature, the molecules of a liquid have a certain average kinetic energy. This is an *average* energy, however. As they bounce around each other, some molecules momentarily have a greater-than-average energy or less-than-average energy. Sometimes a molecule with a lot of energy hits the surface of the liquid and breaks free from the surface. This is the process of evaporation.

(a)   Which would you expect to evaporate more quickly, a warm liquid or a cool one?

_____

(b)   What difference in the molecules causes this difference in evaporation rate?

_____

_ _ _ _ _ _ _ _ _ _ _ _ _ _ _

(a) warm liquid; (b) more fast-moving ones in the warm liquid

6.   Now suppose we get the liquid hot enough so that molecules have enough energy to break free from one another down in the liquid rather than just on the surface. This is

boiling. When a liquid boils, the molecules within the liquid break from one another, forming bubbles of vapor which rise to the surface. To break the molecules free from one another requires much more energy than what is required to break them from fixed positions, and thus the heat of vaporization of water is much greater than the heat of fusion (540 calories per gram rather than 80).

(a)   What is the name of the amount of heat required to break all molecules from a fixed

position in a gram of a substance? _____

(b)   What is the name of the amount of heat required to break all molecules free from

one another in a gram of a substance? _____

(c)   What is happening at the molecular level as a liquid boils? _____

_____

– – – – – – – – – – – – – – –

(a) heat of fusion; (b) heat of vaporization; (c) molecules are breaking free from one another below the surface of the liquid

TRANSFER OF HEAT: CONDUCTION

7.   If you stick the end of a poker into a fire and hold the other end, you will soon feel your end getting hotter. Heat is being transferred from the hot end to the cold end by the method of *conduction.* Picture the molecules in the hot end of the poker in rapid vibration, while the molecules of the cold end are vibrating less rapidly. Since there are forces between the molecules, a rapidly vibrating molecule will transfer some of its vibrational energy to its slower moving neighbor. In this way, the kinetic energy of vibration of the molecules is gradually spread throughout the metal until the molecules at one end have about the same average kinetic energy as the molecules at the other.

Because of the differences between bonding forces in various materials, some materials are better conductors of heat than others. In general, metals are good conductors of heat, and metals which conduct electricity best also conduct heat best. Copper, for example, is a better conductor of both electricity and heat than is aluminum, and silver is better than either. Some materials conduct heat so poorly that we classify them as heat insulators. Wood is only one percent as effective in conducting heat as is aluminum, and is a fair insulator.

(a)   Felt has only about 0.1% the conductivity of aluminum. How many times more

effective an insulator than wood is felt? _____

(b)   Which conducts heat more rapidly, iron or wood? _____

– – – – – – – – – – – – – – –

(a) 10 (since wood is 100 times better than aluminum and felt is 1000 times better than aluminum); (b) iron

8.   Suppose we wish to calculate the amount of heat which will flow through a window pane on a cold winter day. The variables which must be considered are listed below.

(1)   The thickness of the glass. If the glass is thicker, less heat will flow. (We'll call this *d*, for distance.)

(2)   The area of the pane of glass. The greater the area, the more heat flows. (A = area.)
(3)   The difference in temperature between the two surfaces. A greater difference will cause a greater heat flow. (ΔT = temperature difference.)
(4)   The time, t, during which the heat flows.
(5)   The thermal conductivity, k. The table below shows heat conductivities of various materials. The units are $\text{cal} \cdot \text{cm/cm}^2 \cdot \text{s} \cdot \text{C}^\circ$.

The equation relating all of these variables is shown below.

$$H = \frac{k \cdot t \cdot A \cdot \Delta T}{d}$$

The units of H, of course, are calories.

Substance	Thermal conductivity ($\text{cal} \cdot \text{cm/cm}^2 \cdot \text{s} \cdot \text{C}^\circ$)
silver	0.97
copper	0.92
aluminum	0.50
iron	0.16
glass	0.0025
wood	0.0005
felt	0.00004

Now suppose the pane is 15 cm by 20 cm and is 0.3 cm thick. The temperature is 20 Celsius degrees warmer inside than out. How much heat escapes through the window in one hour? _____

– – – – – – – – – – – – – – – –

180,000 calories. $H = \dfrac{0.0025 \cdot 3600 \text{ s} \cdot 300 \text{ cm}^2 \cdot 20 \text{ C}^\circ}{.3 \text{ cm}}$

CONVECTION

9.   Convection is the transfer of heat by transfer of the molecules of the fluid itself. Convection, rather than conduction, is the primary method by which heat is transferred through a fluid. As a substance becomes warmer, it expands. When a solid does this, the molecules must stay in their relative positions, but the expansion of part of a liquid or a gas causes that part of the substance to be less dense than its surroundings, and therefore to rise. (We sometimes say that "heat rises," but what we mean is that hot fluids rise through cooler fluids of the same substance.) You can see convection currents if you place a pan of water on a hot stove and put a drop or two of food coloring in the water.

(a)   What do convection currents do to the process of diffusion? _____

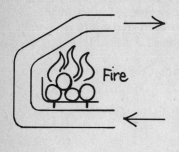

(b)   Pipes such as shown at left are sometimes put in fireplaces to increase the heat output of the fireplace. What method do these pipes use to get heat from the fireplace into the room?

_____

– – – – – – – – – – – – – – – –

(a) speed it up; (b) convection (Heat is transferred from the fire through the walls of the pipes by conduction. Convection then brings the heat out into the room.)

RADIATION

10.   Heat can also be transferred from one place to another by radiation. Every object is constantly absorbing and emitting radiation, but we do not notice it until the object is hot enough to emit thermal radiation to which our skin is sensitive. Such thermal, or infrared, radiation is electromagnetic in nature and travels at the speed of light. Radiant heat is easily felt in front of a fire, a heat lamp, or even near a regular light bulb.

An object which is a good absorber of radiation is also a good radiator. The best absorber (and therefore the best radiator) is a black object. An object with a shiny, silvery surface neither absorbs nor emits radiation very effectively.

(a)   Why are white clothes recommended for hot, sunny locations? _____

_____

(b)   By what method is heat transferred from sun to earth? _____

– – – – – – – – – – – – – – – –

(a)   White clothes do not absorb thermal radiation from the sun as much as do dark colored clothes.
(b)   radiation (The other two methods can play no part out in space because there are no atoms present to conduct or convect heat.)

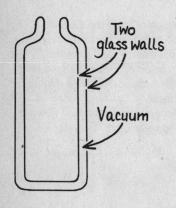

11.   A thermos bottle is constructed so as to reduce heat loss by conduction, convection, and radiation. The figure illustrates the glass within a thermos bottle. Between the two walls is a vacuum. Since a vacuum contains no material, it can neither conduct nor convect heat. Radiation is reduced by silvering the walls of the glass.

(a)   Many old homes are heated by hot water (or steam) "radiators." These work on the same principle as the "radiator" in a car. By what method of heat transfer is most heat carried from such devices? (Hint: Do not be misled by the names.) _____

(b)   Why did the Apollo astronauts wear shiny suits on the moon? _____

_____

(c)   When you touch a lamp post in cold weather, by what method is heat transferred

from your hand? _____

— — — — — — — — — — — — — — — — —

(a)   convection (Air moves across the hot surface of the "radiator," carrying away heat.
       If radiation were the primary method of heat transfer, the devices should be painted
       black!)
(b)   to reflect infrared radiation from the sun
(c)   conduction

SELF-TEST

The questions below will test your understanding of this chapter. Use a separate sheet of
paper for your diagrams or calculations. Compare your answers with the answers provided
following the test. Refer to the table on page 101 at any time.

1.   How many calories of heat must be removed from 5 grams of water at $0°$ Celsius to

      change it into ice at $0°C$? _____

2.   How much heat is needed to change 5 grams of water at $90°C$ to steam at $100°C$?

      _____

3.   Explain, on the molecular level, what the energy is used for when heat is added to

      ice to change it into water. _____

      _____

4.   Explain why an amorphous solid does not have a definite melting point and a heat

      of fusion. _____

      _____

5.   Why is the heat of vaporization of water so much greater than the heat of fusion?

      _____

6.   What are the three methods by which heat is transferred? _____

      _____

7.   Which is the best conductor of heat—iron, silver, or copper? (Use the table on page

      101.) _____

8.   An iron rod has a cross-sectional area of one square centimeter and is one meter long. It is placed so that one end is in a fire at $700°C$ and the other end is kept in water at $50°C$. How much heat flows through the rod in one minute? _____

9.   Gold melts at $1063°C$. Is it an amorphous or crystalline solid? _____

10.   White lines are often painted across asphalt streets in beach areas. Why are the lines cooler to walk on than the asphalt? _____

11.   If a wooden bench and an iron post are both at $-5°C$ (on a cold morning, perhaps), why does the iron post feel cooler to the touch? _____

Answers to Self-Test

If your answers do not agree with those given below, review the frames indicated in parentheses before you go on to the next chapter.

1.   400 calories (5 grams · 80 calories per gram)   (frame 1)

2.   2750 calories. Solution: To change the temperature from $90°C$ to $100°C$ requires 50 calories (calculated from $H = m \cdot c \cdot \Delta T$), and to vaporize 5 grams of water requires 5 · 540, or 2700 calories. Total: 2750 calories.   (frames 2–3)

3.   The energy is used to break the bonds which hold the molecules of the solid ice in their fixed positions.   (frame 4)

4.   In an amorphous solid the individual molecules have bonds of different strength holding them in place. Thus different energies are required to break the bonds and they therefore break at different temperatures. The solid softens as it is heated rather than melting at a definite temperature. Since it does not melt, it has no heat of fusion.   (frame 4)

5.   In vaporization, the molecules must completely break free of one another rather than just loosen their bonds so that they can move around one another. (frame 6)

6.   conduction, convection, and radiation   (frames 7, 9, 10)

7.   silver (The table shows its thermal conductivity to be greatest.)   (frame 8)

8.   62.4 calories. Solution: $H = \dfrac{k \cdot t \cdot A \cdot \Delta T}{d}$

$$= \frac{0.16 \cdot 60 \cdot 1 \cdot 650}{100}$$

$$= 62.4 \quad \text{(frame 8)}$$

9.   crystalline (because it has a definite melting point)   (frame 4)

10.   They do not absorb as much thermal radiation from the sun, so they are cooler. (frame 10)

11.   The iron conducts heat away from our hand much faster than does the wood because iron is a better heat conductor than wood.   (frames 7–8)

PART THREE

The Physics of Waves and Sound

In the modern world we are bombarded with sounds, from the tune of the canary to the rat-a-tat-tat of the jackhammer. The nature and properties of sound form the focus of Part III of this book. Since sound travels from its source in the form of waves, we will first look at wave motion in general, considering various types of waves such as water waves and waves-on-a-string. Then we will proceed to study sound waves—how they are produced, how they travel, and how they are detected in the ear. We'll consider how slight differences in the sound wave allow us to distinguish between the note from a violin and the sound of a car's horn.

The principles of wave motion which will be studied in this section will be used again in Part V, where we will consider light and other electromagnetic waves. Thus a careful study of the next three chapters will be rewarded later.

CHAPTER TEN

Wave Motion

No Prerequisites

Wave motion is basic to an understanding of various aspects of physics, including sound, light, and parts of electricity and magnetism. This chapter will explore wave motion, and the next two will apply the concepts learned to sound.

A study of this "wavy" chapter will enable you to:

- determine the relative periods of simple pendula of different lengths and masses;

- specify the effect of a change of length, mass, or amplitude on the period of a pendulum;

- specify the relationship between frequency and period, and—given one—calculate the other;

- describe a wave in terms of wavelength, amplitude, frequency, and speed;

- use the equation $v = \lambda f$ to determine the velocity of a wave;

- differentiate between transverse and longitudinal waves;

- identify energy as being transmitted by a wave;

- describe the effect of a relative motion between the sender and the receiver of a wave;

- describe a bow wave and state the conditions necessary to achieve one;

- given one of the two, determine the length or period of a simple pendulum (optional).

THE PERIOD OF A PENDULUM

1. The simple pendulum* is one of the simplest mechanical devices in existence. It

---

*All pendula are not "simple" pendula. A simple pendulum has a compact mass hanging from an essentially weightless string. A board swinging from one end or a tire on a short rope are examples of non-simple pendula.

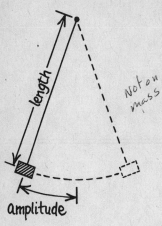

consists of a suspended object (the pendulum "bob") which swings freely. The *period* of a pendulum is the time taken to complete one back-and-forth swing. The period of the pendulum swinging beneath a large grandfather clock is usually two seconds. To a high degree of precision, the period depends only upon the length of the pendulum. (This is true as long as the angle of swing does not exceed 5 or 6 degrees. As the angle—the amplitude—gets larger, the period does also, but the effect is hardly measureable unless large angles are used.) The longer the length, the longer the period.

(a)   If two pendula have the same length, but one has a 2 pound weight on its end while the other has only a 1 pound weight, which will have the

longer period? _____

(b)   A pendulum with a suspending string 2 meters long has a mass of 10 kilograms. Another with a string 4 meters long has a mass of 5 kg. Which will have the longer

period? _____

(c)   If the period of a pendulum is 1 second, how long does it take to go from one end

of its swing to the other? _____

– – – – – – – – – – – – – – – –

(a) the same; (b) 4 meters long; (c) $\frac{1}{2}$ second

Optional   RELATION BETWEEN LENGTH AND PERIOD

2.   The period of a simple pendulum is given by the following equation:

$$t = 2\pi \sqrt{\frac{\ell}{g}}$$

where t is the period, ℓ is the length of the pendulum (measured to the center of mass of the pendulum bob), and g is the acceleration of gravity (about 9.8 m/s^2 or 32 ft/s^2 at the earth's surface—see Chapter 1, frames 6–9).

(a)   By what factor must you increase the length in order to double the period of a

pendulum? _____

(b)   Use the equation to calculate the length needed to obtain a period of 1 second.

(Hint: Start by squaring both sides of the equation.) _____

– – – – – – – – – – – – – – – –

(a)   4 times (since the period depends upon the *square root* of the length)

(b)   0.81 ft. Solution: $t^2 = 4\pi^2 \dfrac{\ell}{g}$

$$\ell = \frac{t^2 g}{4\pi^2}$$

$$= \frac{(1\text{ s})^2 (32\text{ ft/s}^2)}{4(3.14)^2}$$

$$= .81\text{ ft}$$

(A solution using 9.8 m/s² for the acceleration of gravity yields .25 meters as the length. The answers are equivalent.)

FREQUENCY

3.   The *frequency* of a pendulum (or of any regularly repeating motion) is the number of cycles (periods) completed in a unit of time. The period of the grandfather clock pendulum is 2 seconds, which corresponds to a frequency of $\frac{1}{2}$ cycle per second or 30 cycles per minute. The simple relationship between period and frequency is shown below:

$$f = \frac{1}{t}$$

where f is frequency and t is period (abbreviated as it is because period is simply a *time*). The expression $\frac{1}{t}$ is called the reciprocal of t. Frequency is the reciprocal of period.

(a)   What is the frequency of a pendulum that has a 1 second period? _____

(b)   What is the frequency of a pendulum having a period of $\frac{1}{3}$ second? _____

(c)   What is the period of a pendulum having a frequency of 2 cycles/second?

_____

(d)   Which has the greater frequency, a pendulum of length 1 foot or one of length

3 feet? _____

- - - - - - - - - - - - - - - -

(a) 1 cycle/second (or 60 cycles/minute); (b) 3 cycles/second; (c) $\frac{1}{2}$ second;
(d) 1 foot (because it has the smaller period—see frame 1)

WAVES

4.   Suppose a pendulum is arranged so that at one end of its swing it hits the end of a long tank of water. And suppose that somehow the pendulum does not decrease its amplitude, so it continues to hit the tank with each swing. Each time the tank is hit, a small ripple will be sent down the surface of the water. (See the figure on the next page.)

Now if the pendulum is from the grandfather clock, with its 2 second period, these ripples pass a person standing by the tank with the frequency of the pendulum.

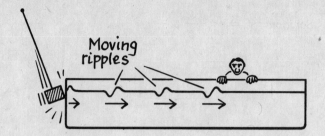

(a)   What is the frequency of the wave caused by the grandfather clock pendulum?

_____

(b)   If the period of a wave is the time between successive high points of the wave as

they pass a given location, what is the period of this wave? _____

— — — — — — — — — — — — — — — — —

(a) $\frac{1}{2}$ wave/second*; (b) 2 seconds

5.   Three more variables enter the picture in determining period and frequency of
waves. The velocity or speed of a wave is defined in exactly the same way as is the speed
of a material object: the distance traveled divided by the time of travel. The *wavelength*
is the distance between two consecutive corresponding points on the wave. In the figure
below, the wavelength of the wave is the distance from X to Y, or from A to C.

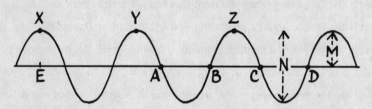

A final variable is *amplitude.* Amplitude of a wave is defined as the distance from
the center, rest position of the wave to the point of maximum displacement. In the figure
the amplitude is the distance M. (Note that it is *not* distance N, which is twice the
amplitude.)

(a)   Which of the variables described in this frame are shown in the figure? _____

_____

(b)   What is the distance from B to D called?_____

(c)   How is the speed of a wave defined? _____

_____

_____

*Notice that the unit has changed from "cycle/second" to "wave/second." Often the
unit of frequency is simply stated as "/second" or "per second."

(d)   What fraction of a wavelength is the distance from A to B?_____

(c)   What is the distance from E to X called?_____

– – – – – – – – – – – – – – – –

(a) wavelength and amplitude; (b) one wavelength; (c) the distance traveled divided by the time of travel; (d) $\frac{1}{2}$; (e) amplitude

RELATIONSHIP BETWEEN THE VARIABLES

6.   To see the relationship between velocity, frequency, and wavelength, imagine yourself on a boat observing waves passing under you. You observe that 30 waves pass the boat per minute, and you discover (with a handy meterstick) that each wave is 2 meters long. Now what is the speed of the waves? Since 30 waves come by each minute and each wave is 2 meters long, the speed is 60 meters/minute. In general, we multiply frequency times wavelength to obtain velocity. Letting the Greek letter lambda (λ) represent wavelength, we have:

$$v = \lambda f.$$

(a)   What is the wavelength in the example above? _____

(b)   What is the frequency? _____

(c)   If the speed of the waves remains the same, but the frequency is doubled, what happens to the wavelength? _____

– – – – – – – – – – – – – – – –

(a) 2 meters; (b) 30 per minute; (c) it is halved

7.   This velocity equation applies to all types of waves, including water waves, sound waves, and light waves. Try a few more applications of $v = \lambda f$.

(a)   Sound waves travel at about 1100 ft/s in air. What is the wavelength of a sound with a frequency of 550 cycles per second? _____

(b)   Suppose the period of some small ripples on water is $\frac{1}{10}$ second, and their wavelength is 2 inches. What is the velocity of the waves?_____

– – – – – – – – – – – – – – – –

(a)   2 ft (which is 1100 ft/s ÷ 550 /s)
(b)   20 in./s (Remember, $f = \frac{1}{t}$.)

TRANSVERSE AND LONGITUDINAL WAVES

8.   Thus far we have spoken of waves that vibrate in a direction perpendicular to the direction the wave is traveling. For example, the primary motion of water is up and down as a surface wave passes. (We are speaking of smooth waves away from the shore; waves breaking on a beach are another matter.) If two girls hold opposite ends of a long spring and one vibrates her end up and down, waves similar to the water waves travel down the

length of the spring. Again, the primary motion of the coils of the spring is up and down although the wave itself is moving horizontally. Such waves are called *transverse waves*.

Now suppose that the two people again hold opposite ends of a spring. (If you want to try this, a Slinky® spring will work best here.) This time one of the people squeezes together a number of coils and releases them. Soon after release the spring will look somewhat like that in the figure below.

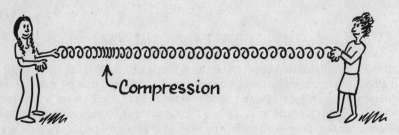

Here the squeezed-up portion (called a *compression*) travels along the spring. The motion of an individual coil as the wave passes is back and forth along the direction of the wave's velocity. Such a wave is called a *longitudinal* wave. If instead of compressing the coils, the person had stretched them apart, a rarefaction would have been produced, and would also have moved along the wave as did the compression. (A *rarefaction* is an area where the coils are stretched out.)

Although transverse and longitudinal waves seem very different in nature, they both obey the basic rules of wave motion, and, of course, the $v = \lambda f$ equation applies to both. We will see that while light acts as a transverse wave, sound waves are longitudinal.

(a)   In what type wave is the vibration perpendicular to the direction of travel of the wave? _____

(b)   What type vibrates parallel to the direction of travel? _____

(c)   What type wave contains compressions and rarefactions? _____

(d)   What type wave is produced when you move one end of a horizontal spring up and down? _____

(e)   What type wave has a wavelength? _____

— — — — — — — — — — — — — — — —

(a) transverse; (b) longitudinal; (c) longitudinal; (d) transverse; (e) both   (Note that the equation applies to both.)

ENERGY TRANSFER

9.   Consider what is moving as a wave moves along a spring. The wave itself moves from one end of the spring to the other, but the individual coils of the spring simply vibrate about a center position. The vibration may be up and down or back and forth, but in neither case do the coils themselves move down the spring. The same idea holds in the case of all wave motion. Waves do not transmit matter, but energy.

We have defined energy (in Chapter 2, frame 6) as the ability to do work and work as the product of force and distance (Chapter 2, frame 1). Now consider a transverse pulse moving along a spring toward the person holding the far end. When the pulse reaches him, he feels a force pulling up (or down) on his hand, and the force will move his hand slightly. Thus it does work on his hand. The person who started the wave pulse put energy into the spring (by exerting a force on it through a distance as he wiggled it), and this energy traveled in the form of a wave to the other end. Unless the person at that end allows his hand to move so as to absorb all of the energy, some of it will be reflected as a wave moving back toward the source.

(a)   What is carried by a wave?_____

(b)   How does this energy transfer occur in the case of longitudinal waves?_____

_____

_____

_____

– – – – – – – – – – – – – –

(a)   energy
(b)   The person starting the wave puts energy in the spring, and it travels to the other end where it exerts a force through a distance on the person there. (The force is back and forth rather than up and down, but the principle is the same.)

THE DOPPLER EFFECT

10.   If pebbles are regularly dropped at the same location into a pool of water, the wave pattern produced will look somewhat like that shown in figure 1 below after four pebbles are dropped. If the location of the drop is changed regularly toward the right so that the waves start from different places, a pattern like that of figure 2 will emerge. Moving the starting point even more rapidly produces the pattern of figure 3.

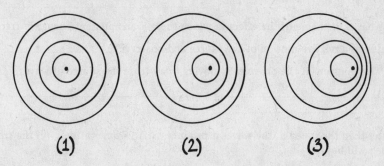

(1)   (2)   (3)

Now consider figure 3. Suppose one person is standing to the left of the place where the pebbles are being dropped (behind the moving pebble-dropper), and one is to the right. Which person sees waves of longer wavelength approaching him?

_____

– – – – – – – – – – – – – –

the one to the left

11. Since the speed of the waves is not affected by their wavelength, a person on the right of the source in figure 3 will see a wave of greater frequency than the frequency at which the pebbles were dropped. And a person to the left of the source in figure 3 will see a lower frequency wave.

(a) Does the apparent *increase* in frequency occur when the source is moving toward the observer or away? _____

(b) When the speed of the source is greater, is the frequency change greater or less?

_____

– – – – – – – – – – – – – – –

(a) toward; (b) greater

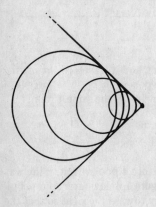

12. If the source continues to move faster until it is moving faster than the speed of the waves themselves, a pattern such as in the figure here results. Notice that along a "V" spreading from the source, the waves overlap and result in an extra-large wave. This is the *bow wave* that you see spreading from a fast-moving boat on a lake.

The effect we have been discussing is called the *Doppler effect*. It also occurs if the observer is moving and the source is stationary. As one moves toward the wave source, the apparent wave frequency is increased.

(a) How fast (relative to the wave speed) must a boat move to produce a bow wave? _____

_____

(b) What is the name of the effect in which there is an apparent change in frequency due to a relative motion of the source and observer? _____

(c) What happens if the observer is moving away from the source of waves? _____

_____

– – – – – – – – – – – – – – –

(a) faster than the speed of the waves it produces; (b) Doppler effect; (c) the frequency observed will be less

SELF-TEST

The questions below will test your understanding of this chapter. Use a separate sheet of paper for your diagrams or calculations. Compare your answers with the answers provided following the test.

1. Draw a simple wave and show what its wavelength is.

2. Locate the amplitude of the wave on the drawing for question 1.

3. You drop a rock in a swimming pool, producing a wave. It reaches the other side, 24 feet away, 4 seconds later. What is the speed of the water wave? _____

4. If you push your hand in and out of the water regularly, you can produce a uniform pattern of waves. Suppose you dip your hand in and out 3 times a second and produce wavelengths of 1.5 feet. What is the speed of the waves?_____

5. The speed of a particular wave is 30 meters/minute. The frequency of the wave is 2 cycles/second. What is its wavelength? _____

6. What is the period of the wave in question 5? _____

7. Distinguish between a transverse and a longitudinal wave. _____
    _____
    _____

8. What is transferred as a wave travels along? _____

9. As a source of waves moves toward you, how does the apparent frequency of the waves compare to when the source was not moving? _____

10. What happens to the apparent frequency as you move away from a source of waves?
    _____

11. Explain what conditions are necessary to form a bow wave and how this wave is formed. _____
    _____

12. *Optional.* What is the period of a pendulum 4 feet long?_____

Answers to Self-Test

If your answers do not agree with those given below, review the frames indicated in parentheses before you go on to the next chapter.

1. Either of the two distances labeled λ (or others this same length) is the wavelength. (frame 5)

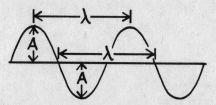

2. Either distance labeled A in the figure is the amplitude.   (frame 5)

3. 6 ft/s (24 ft ÷ 4 s)   (frame 5)

4. 4.5 ft/s ($v = \lambda f = 1.5$ ft $\cdot$ 3/s $= 4.5$ ft/s)   (frame 6)

5. $\frac{1}{4}$ meter or 25 cm (The trick here is that the speed was given in meters per *minute* while frequency was expressed in cycles per *second*. 30 m/min $= \frac{1}{2}$ m/s; $\frac{1}{2}$ m/s ÷ 2/s $= \frac{1}{4}$ m.)   (frame 6)

6. $\frac{1}{2}$ second   (frame 3)

7. In a transverse wave the particles vibrate perpendicular to the direction of travel of the wave, while in a longitudinal wave they vibrate parallel to that direction. (frame 8)

8. energy   (frame 9)

9. The apparent frequency is greater.   (frames 10–11)

10. The apparent frequency is less than when you were stationary.   (frame 12)

11. The source of the waves must be moving faster than the waves themselves move. The waves then "stack up" along a "V" to form the bow wave.   (frame 12)

12. 2.2 seconds. Solution: $t = 2\pi \sqrt{\dfrac{\ell}{g}}$

$$= 2 \cdot 3.14 \sqrt{\dfrac{4 \text{ ft}}{32 \text{ ft/s}^2}}$$

$$= 6.28 \, (.354 \text{ s})$$

$$= 2.2 \text{ s}   \text{(frame 2)}$$

CHAPTER ELEVEN

Sound

Prerequisite: Chapter 10

Sound travels through a material in the form of longitudinal waves. Although we seldom observe directly the wave nature of sound, a consideration of this nature is necessary to understand the sound phenomena which surround us.

When you have completed your study of this chapter, you will be able to:

- describe a sound wave in terms of air molecules;

- identify sound as a longitudinal wave;

- calculate the speed of sound in air at a given temperature;

- calculate the speed of sound using data concerning an echo;

- given either the wavelength or the frequency of a sound, calculate the other quantity;

- relate the speed of sound in a material to the elasticity and density of the material;

- relate intensity to the amplitude of a sound wave;

- calculate relative intensities of two sounds using the inverse square relationship;

- relate resonance to sound waves;

- identify some causes and results of refraction of sound waves;

- compare the speed of sound in air to its speed in other materials;

- identify the frequency range of audible sound;

- calculate the sound intensity in SI units given the decibel level (optional).

SOUND WAVES

1.   Part 1 of the figure below is a representation of a sound wave. Note that condensations and rarefactions follow one another through the material as the wave moves along. Since the material through which we normally experience the motion of sound is air, the dots here represent millions of molecules of air. At point X the molecules are in the center of a compression. As the wave moves along, that point will change to a rarefaction, then back to a compression, and so on.

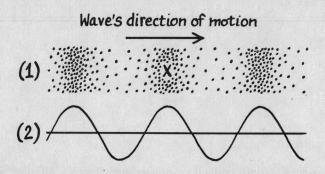

Wave's direction of motion

(1)

(2)

In part 2 of the figure the sound wave is represented graphically. Note that where there is a compression of the wave, the curve is at a maximum above center, and where there is a rarefaction it is at a maximum below center. Since the air is under greater pressure at a compression than it is normally, and under less pressure at a rarefaction, the graph can be taken to represent the pressure of the air.

The fact that part 2 of the figure appears to represent a transverse wave can cause confusion. Sound is not a transverse wave and we are not calling it one. Remember that the curve is a graph of the pressure (or density), and is not meant to be a drawing of the wave itself.

(a)   What corresponds to the highest point of amplitude, compression or rarefaction?

_____

(b)   Can you tell from this frame why sound doesn't travel in a vacuum? Explain.

_____

– – – – – – – – – – – – – – – –

(a) compression; (b) there is nothing to be compressed in a vacuum

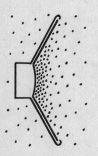

2.   The figure here represents the speaker of a radio or stereo. If you touch such a speaker (gently!) while it is producing sound, you will feel a vibration. It is this in-and-out motion of the speaker which sets up the sound wave in the air. In the figure the speaker cone has just moved to the right. This pushes the molecules of the air together in front of the speaker, causing a compression. Just as it does in a spring, the compression in the air moves away from the source, here the speaker. Meanwhile the speaker cone is moving toward the left and producing a rarefaction in

front of it. This rarefaction follows the compression through the air. Then the cone moves toward the right, producing another compression, and the cycle continues. A sound wave moves away from the speaker through the air. The air itself is not moving away from the speaker; the molecules of air are vibrating as the *wave* moves toward the right.

Your eardrum is somewhat similar to a speaker cone, but it is used as a receiver. When a compression hits the eardrum, the higher pressure causes the eardrum to move inward. Then the following rarefaction causes it to move outward. This pressure wave is changed to electrical signals by the inner ear and then is transmitted to the brain for interpretation.

(a)   How do molecules of air vibrate in a sound wave, perpendicular or parallel to the

       direction of the wave? _____

(b)   What always follows a compression in a sound wave? _____

(c)   What is transmitted in a sound wave, air molecules or energy? _____

- - - - - - - - - - - - - - -

(a) parallel; (b) rarefaction; (c) energy

THE SPEED OF SOUND

3.   The speed of sound in air is relatively easy to measure. Suppose you stand in an open field at a distance from the flat side of a large barn and clap your hands. You will hear an echo when the sound is reflected from the barn back to you. If you are 550 feet from the reflector, the echo will be heard about one second after the clap. Remembering that the sound travels to the barn and back to you in the one second, calculate the speed

of sound. _____

- - - - - - - - - - - - - - -

1100 ft/s (about 750 miles per hour, $\frac{1}{5}$ mile per second, or about 335 meters per second)

4.   The speed of sound in air actually depends upon the temperature and humidity of the air. The 1100 ft/s given above is the speed in dry air at a temperature of 5°C (41°F). For each Celsius degree rise above 5° the speed increases about 2 ft/s.

The speed of sound in water is about 4700 ft/s, more than four times greater than in air. In steel, the speed is about 16,500 ft/s. In general, the speed of sound is greater in more elastic materials and less in more dense materials. (Although steel is more dense than water, it is far more elastic, resulting in the much greater speed.)

(a)   Seventy-two degrees Fahrenheit is about 22°C. What is the speed of sound at this

       temperature? _____

(b)   How far will sound travel in air at 72°F in 3 seconds? _____

(c)   How far will sound travel in water in 3 seconds? _____

(d)   Does sound travel faster in steel or in water? _____

- - - - - - - - - - - - - - -

(a)   1134 ft/s $(22 - 5 = 17; 17 \cdot 2 = 34; 34 + 1100 = 1134)$; (b) 3402 ft;
(c) 14,100 ft; (d) steel

FREQUENCY AND WAVELENGTH RANGE

5.   The human ear is sensitive to frequencies ranging from about 20 cycles/s to 20,000 /s. Waves with frequencies greater than 20,000 /s are said to be *ultrasonic* waves. Using the relationship between speed, frequency, and wavelength from the last chapter ($v = \lambda f$), calculate the range of wavelengths of sound in air. (Use 1100 ft/s as the speed of sound.)

(a)   What is the wavelength of a 20 cycle/s sound? _____

(b)   What is the wavelength of a 20,000 cycle/s sound? _____

- - - - - - - - - - - - - - -

(a) 55 ft (1100/20); (b) .055 ft, which is .66 inch

(See Chapter 10, frame 7 for a review of this relationship.)

INTENSITY

6.   The *intensity* of a sound wave is defined as the amount of sound power passing through a unit area. In Chapter 2 power was defined as energy divided by time. In SI units, intensity is the amount of energy passing through a square meter of area in a second. Watts/m^2 is the SI unit of intensity. The difference between the pressure (or density) of the compression and the rarefaction determines the intensity of the wave.

(a)   On a graphical representation of sound waves, will the intensity be shown by the amplitude, the wavelength, or the frequency? _____

(b)   Which of the sound waves in the figure has a greater intensity? _____

- - - - - - - - - - - - - - -

(a) amplitude; (b) Y

7.   The actual amount of energy carried by a sound wave is small. It would take thousands of people shouting to produce enough sound energy to light a regular light bulb, even if all of the sound energy could be changed to electrical energy. The pressure difference between a compression and a rarefaction in air is only about 10^{-5} pound/in.^{2} for conversational sound levels. (Recall that regular atmospheric pressure is about 15 pounds/in.^{2}.)

Would a sound wave with a pressure difference of 10^{-4} pound/in.^{2} be more or less intense than conversational sound level? _____

- - - - - - - - - - - - - - -

more (See Appendix I for powers-of-ten notation.)

THE DECIBEL SCALE

8.   The metric unit of sound intensity is watts/m^2. While this unit provides us with a standard which is used in all fields of physics, intensity of sound has historically been expressed in decibels (abbreviated db). The decibel scale is a compressed scale of measurement. In this system, the least intense sound that can be heard by a good ear has an intensity of zero decibels. A sound carrying ten times as much energy is rated as 10 decibels, a sound with 100 times as much energy as the zero db sound is rated at 20 db, and 1000 times as much energy at 30 db.

On the decibel scale a whisper measures about 20 db, and ordinary conversation is about 60 db. Sound becomes painful at about 120 db, the "threshold of pain." At a distance of about 100 yards, a rocket engine has an intensity of about 180 db, well above the threshold of pain.

(a)   How many decibels is 10,000 times the energy of 0 db? _____

(b)   Is this above the threshold of pain? _____

(c)   Rock concerts have been measured at 125 db. Is this above the threshold of pain?

_____

_ _ _ _ _ _ _ _ _ _ _ _ _ _ _ _

(a) 40 db; (b) no, not nearly; (c) yes (Prolonged sound at this level causes permanent hearing damage.)

Optional   INTENSITY CALCULATIONS

9.   A zero decibel sound has an intensity of 10^{-12} watt/m^2. A 10 db sound thus has an intensity of 10^{-11} watt/m^2. Recall that an increase of 10 db corresponds to a tenfold energy increase.

(a)   What is the intensity (in SI units) of a 20 db sound? _____   _____

(b)   How much more energy than 0 db does 20 db carry? _____

(c)   How many times more intense is an 80 db sound than a 50 db sound?

_____

(d)   What is the intensity in watts/m^2 of a 70 db sound? _____

_ _ _ _ _ _ _ _ _ _ _ _ _ _ _ _

(a) 10^{-10} watt/m^2; (b) 100 times; (c) 1000 times, because the difference in decibels is 30; (d) 10^{-5} watt/m^2 (An increase of 70 db multiplies the intensity by 10^7 times. $10^7 \cdot 10^{-12} = 10^{-5}$)

DECREASE OF INTENSITY WITH DISTANCE

10.   When a pebble is dropped into a calm lake, a ripple spreads over the surface of the water in an ever-enlarging circle. Consider the energy carried by the wave. As the circle gets larger and larger, the energy becomes more spread out, so that by the time the ripple arrives at a distant shore the ripple is barely perceptible. Except for the energy which is

changed into heat due to the stirring of the water, the original amount of energy is still in the ripple, but it has spread over such a large circle that the energy contained in a short section of the circle is very little.

When you ring a bell, the sound waves produced spread out in much the same way that the water wave did, except that the spreading is now in three dimensions rather than in just two. Instead of an ever-enlarging circle, we have an ever-enlarging sphere. The area of a sphere is given by the formula:

$$A = 4\pi r^2$$

where r is the radius of the sphere. Intensity is defined as the power passing through a unit area. Since the area of a sphere depends upon the square of the radius (r^2), the energy carried by the wave spreads out inversely proportional to the square of the radius. Thus intensity, I, depends upon the reciprocal of the radius squared, or:

$$I \sim \frac{1}{r^2}.$$

Assume that the intensity of sound 10 feet from a certain source is 1.

(a)   Using the formula, how intense will the sound be at 20 feet (twice as far away as 10 feet)? _____

(b)   How intense will it be at 30 feet? _____

– – – – – – – – – – – – – –

(a) $\frac{1}{4}$ $(1/2^2)$; (b) $\frac{1}{9}$

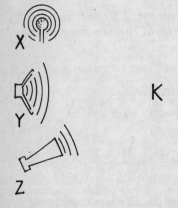

11.   Actually, we have neglected the fact that some sound energy is absorbed by the air. Thus the actual sound intensity is reduced even more quickly than indicated in frame 10. The inverse square effect due to the spreading of the energy is the major cause of the decrease in intensity, however.

One way to reduce the effect of the spreading is to "funnel" the sound in a particular direction by using a device such as a megaphone. This causes most of the energy to be sent in the desired direction. Three sources of sound are shown in the figure. Each emits the same power. Which produces the loudest

sound at K? _____

– – – – – – – – – – – – – –

source Z

RESONANCE

12.   Every object has a natural frequency at which it will vibrate. Kick the wall beside you and you will probably hear a low frequency sound. If you drop a spoon, it will vibrate with a higher frequency. You may have had the experience of hearing a table or cabinet vibrate when a hi fi speaker (playing loudly) is placed on or near it. This is an

example of *resonance*. As a sound wave travels through the air it carries energy with it. This energy is what causes our eardrums to vibrate when struck by the wave. In the same manner, a larger, less sensitive object also bends slightly with each compression and rarefaction which hits it. Normally this deformation is so slight that it is not noticeable, but if the frequency of the sound wave just matches the natural frequency of the object struck, the object will pick up the vibration—it will *resonate*.

It is said that some singers can shatter a crystal goblet with their voice.

(a)   What is it about a loud sound which enables it to break the goblet when a quieter

sound won't? _____

(b)   Why will the goblet break due to sound of one frequency but not sound of a differ-

ent frequency? _____

— — — — — — — — — — — — — — — —

(a)   The louder sound carries more energy. (Saying that the louder sound has a greater amplitude is an equivalent answer.)
(b)   The frequency of the sound must match the natural frequency of the goblet.

REFRACTION

13.   We said earlier that sound travels faster in warm air than in cold. Consider what happens when air near the ground is cooler than air above the ground. As sound waves travel away from a source, those going up into the warmer air start to travel faster. This results in the waves being bent from their original directions, as shown in the figure below. The effect is that the waves tend to be bent back down toward the ground. Such a bending of waves is called *refraction*. Refraction occurs whenever waves pass at an angle into an area where they travel at a different speed.

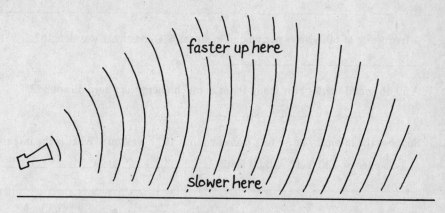

faster up here

slower here

(a)   Indicate which of these might be a factor in refraction: _____ air temperature

_____ intensity   _____ material wave passes through

(b)   Would you expect refraction to affect the wavelength or the frequency of the

sound? _____

(c)   What happens to the direction of a wave when it is refracted? _____

(d)   What would happen to a sound wave if the air near the ground were warmer than above (so that the sound traveled faster near the ground)? _____
_____

– – – – – – – – – – – – – – – –

(a) air temperature and material; (b) wavelength; (c) it changes; (d) it would bend upward

SELF-TEST

The questions below will test your understanding of this chapter. Use a separate sheet of paper for your diagrams or calculations. Compare your answers with the answers provided following the test.

1.   Are sound waves transverse or longitudinal? _____

2.   How do the molecules in a compression differ from those in a rarefaction?
_____

3.   What is the speed of sound in air at a temperature of 18°C? _____

4.   What would you predict concerning the speed of sound in a metal which has the same density as steel but is less elastic than steel? _____

5.   What are the lower and upper limits of frequency to which the human ear is sensitive? _____

6.   A frequency of 500 cycles per second in air (at 5°C) has what wavelength?
_____

7.   A 50 db sound carries how many times as much energy as a zero db sound?
_____

8.   Suppose the intensity at 50 feet from a siren is 10^{-5} watt/m^2. What is the intensity at a distance of 100 feet? At 200 feet? _____
_____

9.   If one moves twice as close to a source of sound, how does the intensity reaching him change? (Assume that there is negligible reflection of sound from surroundings.)
_____

10.   What must be different about how sound travels in two materials in order for refraction to occur when sound goes from one material to the other? _____
_____

11.   *Optional.* What is the intensity in decibels of a sound with an intensity of 10^{-4}

      watt/m^2? _____

12.   *Optional.* What is the intensity (in watts/m^2) of a sound of 50 db? _____

13.   *Optional.* How many times more intense is a 70 db sound than a 40 db sound?

      _____

Answers to Self-Test

If your answers do not agree with those given below, review the frames indicated in parentheses before you go on to the next chapter.

1.   longitudinal   (frame 1)

2.   They are closer together (and therefore under more pressure) in a compression. (frames 1–2)

3.   1126 ft/s. Solution: $18°$ is $13°$ higher than $5°$. At 2 ft/s increase per degree, this amounts to a $26°$ rise above the 1100 ft/s speed at $5°$C.   (frame 4)

4.   It would be less.   (frame 4)

5.   20 and 20,000 cycles per second   (frame 5)

6.   2.2 ft or 0.67 meter (This is obtained by dividing the speed—taken as 1100 ft/s or 335 m/s—by the frequency.)   (frame 5)

7.   100,000 times as much (This is 10 times as much for each rise of 10 db.) (frame 8)

8.   At 100 ft: $0.25 \cdot 10^{-5}$ watt/m^2 (because at twice the distance the intensity is reduced to $\frac{1}{4}$ as much)

     At 200 ft: $0.0625 \cdot 10^{-5}$ watt/m^2 (which is better written as $6.25 \cdot 10^{-7}$ watt/m^2. 200 ft is 4 times farther away, which reduces intensity to $\frac{1}{16}$ as much.) (frame 10)

9.   It gets four times as great.   (frame 10)

10.   The speed of sound must be different in the two materials.   (frame 13)

11.   80 db (10^{-4} watt/m^2 is 10^{-12} watt/m^2 $\times$ 10^8. Thus the zero db level sound has been increased by a factor of 10^8. This corresponds to 80 db.)   (frame 9)

12. 10^{-7} watt/m^2 (50 db is 10^5 times more intense than 10^{-12} watt/m^2—the zero db level.)   (frame 9)

13.   1000 times more intense ($70 - 40 = 30$. Thus the intensity has been increased by a factor of 10^3.)   (frame 9)

CHAPTER TWELVE

Diffraction, Interference, and Music

Prerequisites: Chapters 10 and 11

In this chapter we will look further into the wave nature of sound and consider phenomena which require us to delve into what happens when two sound waves overlap.

This chapter will enable you to:

- relate loudness to intensity;

- relate pitch to frequency;

- differentiate between refraction and diffraction;

- specify the cause of sound wave interference;

- differentiate between constructive and destructive interference;

- determine the beat frequency of two sound sources;

- specify what determines the quality of sound;

- relate a fundamental frequency to its overtones in terms of frequency;

- differentiate between standing and traveling waves;

- identify nodes and antinodes in a standing wave;

- relate the Doppler effect to sound waves;

- describe the conditions necessary to produce a sonic boom and explain what causes it;

- given a drawing of a standing wave, specify the number of wavelengths and which overtone is represented;

- describe the conditions necessary for beats to be produced.

LOUDNESS

1. In our everyday language we may use the term loudness to mean intensity. Intensity, however, refers to a measurable, physical attribute of a sound wave, while loudness concerns a sensation caused by a sound wave. You may perceive the same sound to be louder than I do, but intensity is measurable and not subject to individual interpretation. In general, however, the more intense the sound, the louder it seems. But the relationship is very complicated, involving not only intensity, but also (to a lesser degree) frequency.

For one sound to appear twice as loud as another, its intensity must be—not twice as much—but eight to ten times as much. Because of this logarithmic effect of loudness, our ears are sensitive to a very great range of sound intensities. We detect sound energies as weak as 10^{-12} watt/m^2, and we can hear without pain sounds with intensities up to about 1 watt/m^2 (although extended exposure to intensities above 10^{-3} watt/m^2—90 db—can cause permanent hearing loss).

Suppose you know that the sound of a whisper is 20 db—or has intensity of 10^{-10} watt/m^2. You hear another sound that seems to be *twice* as loud as a whisper. What would you guess its intensity to be: twice 10^{-10}, one half 10^{-10}, or ten times 10^{-10}?

_____

$10 \cdot 10^{-10}$ (which is 10^{-9})

DIFFRACTION OF SOUND

2. *Diffraction* is defined as the spreading out of a wave as it goes around a corner. It is obvious that sound waves do spread around corners, for you can hear a person in the next room even though he may be out of sight. (Some sound actually passes through the wall, but most of the sound you hear comes through the door opening.)

Notice that the bending which occurs in diffraction differs from that in refraction. When waves are refracted they no longer continue in the same direction, but in diffraction only part of the wave goes in a new direction. Consider sound going through a door from one room to another. As shown in the figure at left, the sound continues straight into the second room as well as bending around the door opening.

Label the drawings below as diffraction or refraction.

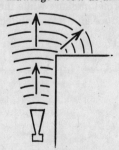

_____

The left one is diffraction and the right one is refraction. (See Chapter 11, frame 13)

INTERFERENCE

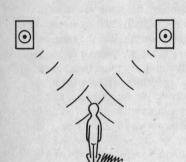

3.   Consider a person listening to a constant-frequency sound from two speakers as shown in the figure here. Both speakers are placed at the same distance from the listener, and both are connected to the amplifier so that when one is producing a condensation the other is also. When these conditions are met, the person is struck by condensations from both speakers at the same time, then by rarefactions from both speakers. The waves are said to be striking him *in phase.* Both speakers cause a change in pressure on the listener's eardrum at the same time.

(a)   With two sound waves in phase, is the combined effect of a rarefaction more, less, or the same pressure than from a single source? _____

(b)   How would you expect in-phase sound to affect the amplitude of a sound wave?

_____

(c)   What effect on the loudness would you expect due to two sound waves being in phase? _____

– – – – – – – – – – – – – – – –

(a) less pressure (and a condensation would have more pressure due to the combined waves); (b) the sound wave which results would have greater amplitude; (c) the sound would be louder

4.   Now consider the case shown in the figure in this frame. Here one speaker is located exactly one half wavelength further from the listener than the other speaker. Now when a condensation from the left speaker reaches his ear, a rarefaction is reaching it from the right speaker. Likewise, when a rarefaction reaches the ear from the left speaker, a condensation reaches it from the other. In each case the condensation alone would produce an increase in pressure over normal pressure while the rarefaction alone would produce an equivalent decrease in

pressure. What would you expect as the overall effect? _____

_____

– – – – – – – – – – – – – – – –

they would cancel each other out

5.   When two sound waves are in phase, increasing loudness or intensity of sound, we say that we have *constructive interference*. When the waves cancel one another out, resulting (in the ideal case) in no sound being heard by the listener, we call it *destructive interference*. (In practice, complete cancellation is difficult to achieve, mainly because of sound bouncing back from nearby surfaces.)

(a)   What is interference in sound waves? _____

_____

(b)   How does constructive interference differ from destructive interference?

_____

— — — — — — — — — — — — — —

(a) one wave affecting or changing another when the two overlap; (b) constructive interference increases intensity while destructive interference decreases it

6.   Recall that sound waves may be represented by pressure graphs. Figure A at the left represents the sound wave from some source, and B represents another sound wave at the same location as wave A. Figure C shows the combined effect—the waves added together.

(a)   Does Figure C represent constructive or destructive interference? _____

(b)   The lower three graphs represent two more waves (D and E) and their total (F). Are D and E in phase or out of phase? _____

(c)   Do D and E interfere constructively or destructively? _____

— — — — — — — — — — — — — —

(a) constructive; (b) out of phase; (c) destructively

BEATS

7.   Figure 12.1 shows two side-by-side speakers emitting sound waves which differ slightly in frequency. The speaker at top has the longer wavelength and therefore the lesser frequency. The waves are shown traveling beside one another, but in actual practice they overlap, causing interference. Where the two waves meet with condensation on condensation, constructive interference results, but where condensation lands on rarefaction, destructive interference results.

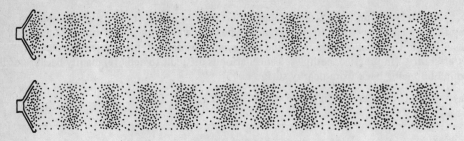

Figure 12.1

In waves A and B below, the two waves are shown in graphical form. The resultant wave is then shown as wave C. Note that there are areas of wave C where the amplitude is greater than the amplitude of either wave alone, but at other places the waves are cancelled out.

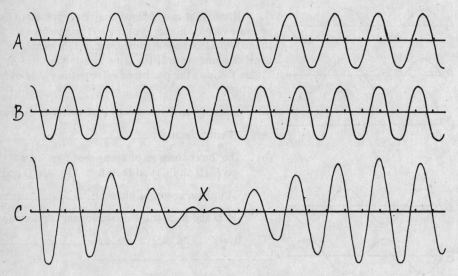

Now picture this resultant wave moving toward a listener at the right. The listener hears sound when the large amplitude waves hit his ear, but the sound dies down when the points of destructive interference arrive. This effect of rapidly changing loudness of sound produces *beats*, and it can be heard whenever two sounds of nearly the same frequency are sounded together.

(a)   Is point X in wave C a place of constructive or destructive interference?

_____

(b)   Is the phenomenon of beats the result of constructive interference, destructive interference, or both? _____

- - - - - - - - - - - - - -

(a) destructive; (b) both

8.   The *beat frequency* is the frequency with which the cycle of loud and no-so-loud sound occurs. This beat frequency is equal to the difference between the frequencies of the two sounds producing the beats. Thus if one source vibrates with a frequency of 500 /s and the other vibrates at 503 /s, the beat frequency is 3 /s. A piano tuner uses a sound source of known frequency (normally a tuning fork) and adjusts the appropriate piano string until the beats between the two are eliminated.

If the frequencies in Figure 12.1 were 420 /s and 430 /s, what would be the beat

frequency heard? _____

— — — — — — — — — — — — — — — —

10 /s

PITCH

9.   Pitch is the subjective impression which distinguishes sounds by differences in tone. We hear a higher pitch when we strike a key at the right end of a piano keyboard than when we strike one at the left. Pitch depends primarily on frequency, a higher frequency sound generally being heard as a higher pitch, but while frequency is a physical, measurable quantity, the pitch of a sound is a hearing sensation and is not easily measured.

(a)   Is pitch more like loudness or intensity in terms of measurability? _____

(b)   Would a sound with a frequency of 500 /s or 400 /s be perceived as a higher pitch?

_____

— — — — — — — — — — — — — — — —

(a) loudness; (b) 500 /s

QUALITY

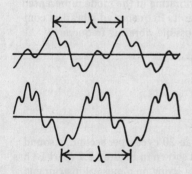

10.   Thus far we have considered only sounds which can be represented graphically as smooth, regular waves.* In actual practice, few sounds are so "pure." A tuning fork, when struck correctly, produces a pure tone, and a whistle made by pursing the lips is often a pure tone, but most sounds we hear are combinations of a number of frequencies. The graphs at the left show what the same note played on two different musical instruments may look like. Note that the wavelength—the distance between complete repeating cycles—is the same for both. (The wavelength is measured for entire repeating cycles. Don't be distracted by the variations.) Even though the basic wavelengths of the two sounds are the same, we can tell the sounds apart when we hear them. We say they differ in *quality*.

(a)   How would you expect the two sounds represented by the graphs to compare in

frequency? _____

_____

*Often called sine waves or sine curves.

(b)   How would you expect them to compare in pitch? _____

(c)   The two curves do differ in appearance. Would you expect them to sound identical?

_____

– – – – – – – – – – – – – – – –

(a) identical (because the wavelengths are identical); (b) same; (c) no (The quality is what enables you to tell a piano from a guitar.)

11.   The quality of a tone is determined by the number and relative intensities of its overtones. To consider how these frequencies are produced, we will consider vibrating strings (which can produce the waves). When a string is fixed at its two ends it can be made to vibrate in a number of different ways.

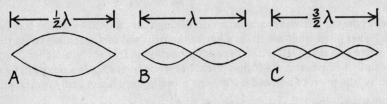

Figure 12.2

Figure 12.2 (part A) shows the simplest vibration. This is called the *fundamental* mode of vibration. If you ever tied one end of a rope to something and shook the other end, you know that you can make the rope vibrate as shown in part A of the figure, and that if you shook your end faster you could produce a vibration such as in B. To get a vibration mode like B, you vibrate the rope with twice the frequency of A (which represents the lowest vibrating frequency). Figure C shows three segments in the rope. Its frequency of vibration is three times that of A.

If a string is vibrating in the mode represented at the left, how does its frequency of vibration compare to its lowest possible vibrating frequency?

_____

– – – – – – – – – – – – – – –

five times as much

12.   If the vibration frequency of a string is greater than 20 cycles per second, a sound will be produced. The sound produced by the vibration shown in Figure 12.2 (part B) has twice the frequency of that in A. In practice, however, a string on a musical instrument vibrates in a number of different modes at the same time. The lowest frequency of vibration, the fundamental, determines the pitch of the sound we hear. The next highest frequency, produced by the vibration of part B, is called the *first overtone*. The third mode shown (C) is the second overtone and has a frequency three times as great as the fundamental. The resulting sound wave which we hear is thus the product of interference of a number of waves.

(a)   What determines the quality of a tone? _____

(b)   How does the frequency of the third overtone compare to that of the fundamental?

_____

_ _ _ _ _ _ _ _ _ _ _ _ _ _ _ _

(a) overtones (the number and relative intensity); (b) four times as much

STANDING WAVES IN A STRING

13.   When a string is vibrated as shown in Figure 12.2, instead of waves appearing to move down the string and back, the waves appear to be standing in one place. This type vibration is appropriately called a *standing wave*. In part A the string was vibrating in its fundamental mode, and the distance from one end of the string to the other was half the wavelength of the transverse wave in the string. In B, the distance between ends was one wavelength. Since the speed of the waves in the string does not depend upon the wavelength, we can conclude that the frequency in B is twice as much as in A.

In C, the distance between ends was $1\frac{1}{2}$ wavelengths and the frequency was three times as much as in the fundamental vibration. In general, when a string is fixed at both ends, one can set up standing waves in it of $\frac{1}{2}$, 1, $1\frac{1}{2}$, 2, $2\frac{1}{2}$, . . . wavelengths. This is usually written in equation form as:

$$d = \frac{n\lambda}{2}. \quad n = 1, 2, 3, 4, \ldots$$

where d is the length of the string and λ is the wavelength. (An equation such as this is really an infinite number of equations, for there is an equation for each value of n.)

(a)   How many wavelengths long is the string when $n = 5$? _____

(b)   How does a standing wave differ from a moving one?_____

_____

_ _ _ _ _ _ _ _ _ _ _ _ _ _ _ _

(a) $2\frac{1}{2}$; (b) The waves just vibrate back and forth; they don't travel along the string.

14.   The figure below represents a standing wave $2\frac{1}{2}$ wavelengths long. The points where the string is not in motion, points A, C, and E for example, are called *nodes*. The points midway between the nodes, where the vibration is at its maximum, are called *antinodes*.

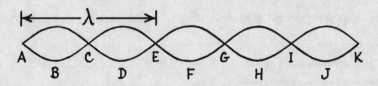

(a)   Consecutive antinodes are separated by how many wavelengths?_____

(b)   What is the distance between a node and the closest antinode? _____

_ _ _ _ _ _ _ _ _ _ _ _ _ _ _ _

(a) $\frac{1}{2}$; (b) $\frac{1}{4}$ wavelength

15. The standing waves shown thus far all have had nodes at their ends. Although it is not easy to set up in practice, one can imagine a standing wave in a string in which the

qualifying conditions are that it has a node at one end and an antinode at the other. Such a wave is shown at the left.

(a)   The length of the string here is how many wavelengths?_____

(b)   How many wavelengths long is the string when it is vibrating at its lowest possible frequency under these qualifying conditions? _____

(c)   The equation for the list of possible wavelengths for this type is:

$$d = (2n - 1)\frac{\lambda}{4}. \quad n = 1, 2, 3, \dots$$

What value does this give for d when $n = 1$?_____

(a) $1\frac{1}{4}$; (b) $\frac{1}{4}$  ; (c) $\frac{1}{4}\lambda$ (You might substitute some values of n in the equation to satisfy yourself that it works. For $n = 3$ we get the situation in the figure.)

STANDING WAVES IN A GAS

16.   The reason that we have spent so much time on standing waves in a string is that they are much more easily visualized than the longitudinal standing waves of sound. Such waves occur when sound of the right frequency travels down a tube and is reflected back toward the open end. Although the waves in this case are longitudinal, we represent them

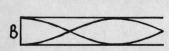

graphically with drawings which look just like those of the transverse standing waves in a string.

A standing wave can be set up in a tube which has one end closed if there is an antinode at the closed end and a node at the open end.* Thus the length of the tube will be $\frac{1}{4}$ wavelength for the lowest frequency wave. Figure A at the left represents such a wave. The next highest frequency which will cause a standing wave in this tube is shown in B.

(a)   How many wavelength(s) long is the wave in B? _____

(b)   Does wave B represent the fundamental, first overtone, or second overtone?

_____

(c)   How many nodes and antinodes are shown in B? _____

(a) $\frac{3}{4}$; (b) first overtone; (c) two of each

------

*If you consult other texts, you'll find that some reverse the positions of node and antinode. The wave they are considering represents the displacement of molecules in the wave. We are representing the pressure.

17.   A longer tube produces a lower fundamental frequency, because the wavelength is longer. This is the principle of wind instruments and is carried to the ultimate in the pipe organ.

One may also set up standing waves in a tube which is open at both ends, because some sound will be reflected back from the open end. In this case a node will be at both ends. This situation is similar to that of a string which is held at both ends.

(a)   At which length will the fundamental frequency occur for a tube open at both ends?

_____

(b)   How does this compare to the closed end pipe? _____

————————————————————

(a)   when the length is $\frac{1}{2}$ wavelength

(b)   the length of the pipe is $\frac{1}{4}$ wavelength for the closed pipe

THE DOPPLER EFFECT IN SOUND

18.   In Chapter 10 (frame 10) we saw that when a source of sound moves away from the observer, the frequency at which the waves pass the observer is less than when the source is stationary. As the source moves toward the observer, the frequency is greater. This phenomenon occurs also in the case of sound. Since a high frequency corresponds to a high pitch, and a low frequency to a low pitch, you can see that the Doppler effect explains the apparent change in pitch of the sound of a race car as it rushes by.

It is important to see what the Doppler effect is *not* saying. First, it is not concerned with the loudness of the sound. Sure, the sound becomes louder as the car approaches, but this is nothing more than the inverse square effect discussed in Chapter 11, frame 10. Also, the Doppler effect does not say that the frequency increases as the car comes closer. It says that the frequency is *constantly* higher all of the time as the car approaches, and constantly lower as the car recedes.

(a)   What explains the higher pitch of a police car siren when it is approaching you?

_____

(b)   What explains the fact that a jack hammer sounds louder when you are next to it

than when you are far away? _____

————————————————————

(a) Doppler effect; (b) inverse square law

THE SONIC BOOM

19.   When a wave source on water moves at a speed greater than the speed of the water waves, a bow wave is produced. We can see now that the bow wave is in fact the result of constructive interference of waves as the peaks pile one on another. Again, a similar phenomenon occurs in the case of sound.

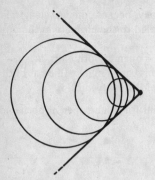

The figure at the left is similar to the figure in Chapter 10, frame 12, but here the lines represent condensations of a sound wave. Notice that they pile up just as did the water waves. But since the sound waves spread not only in two dimensions, but up and down as well, the figure formed is not simply a "V," but a cone. A cone of highly compressed air spreads from any object which is moving faster than the speed of sound. When this compression reaches someone, he experiences a sudden, loud sound, a *sonic boom*. The pressure amplitude of sonic booms may be sufficient to break windows or crack plaster.

Note that a sonic boom is produced all the time that a plane is flying at a speed greater than the speed of sound, and *not* just at the instant that the plane goes from slower than the speed of sound to faster than the speed of sound (called "breaking the sound barrier" because in order to do this, the plane must push through the wall of high pressure in front of it).

(a)   How fast must an object go (at 5°C) to create a sonic boom? _____

(b)   Can a sonic boom be created by an object moving at a speed of 2000 ft/s? _____

(c)   Try to explain how a sonic boom can cause actual damage. _____

_____

_ _ _ _ _ _ _ _ _ _ _ _ _ _ _

(a) 1100 ft/s (the speed of sound—Chapter 11, frame 4); (b) yes; (c) the sudden pressure change does it

SELF-TEST

The questions below will test your understanding of this chapter. Use a separate sheet of paper for your diagrams or calculations. Compare your answers with the answers provided following the test.

1.   How does loudness differ from intensity of sound? _____

_____

2.   What is meant when it is said that two sound waves are "in phase"? _____

_____

3.   What is the difference between constructive and destructive interference? _____

_____

_____

4.   What conditions are necessary to produce beats? _____

_____

5.   If two tuning forks—one 723 cycles/s and the other 727 /s—are sounded together, what beat frequency is heard? _____

6.   What is the difference between pitch and frequency? _____

_____

7.   What determines the quality of a note? _____

_____

8.   How are we able to distinguish between a note played on a flute and the same note played on a clarinet? _____

9.   If a string is held at both ends, how would it appear when vibrating at its fundamental frequency? _____

10.   The figure below represents a string in vibration. Its frequency of vibration is how many times as great as its fundamental frequency (when held the same way)? What overtone does this vibration represent? _____

11.   What is a common phenomenon that is explained by the Doppler effect in sound?

_____

12.   What conditions are necessary to produce a sonic boom? _____

_____

13.   How many wavelengths long is the string in the figure of question 10? _____

14.   A long whip can be made to "crack" by making its end to move at a speed greater than the speed of sound. What causes the "crack" or "pop"? _____

_____

Answers to Self-Test

If your answers do not agree with those given below, review the frames indicated in parentheses before you go on to the next chapter.

1.   Loudness is the physiological sensation while intensity is a direct measure of the power per square meter of the wave.   (frame 1)

2.   "In phase" means that when one wave is at a condensation, the other is also. (frame 3)

3.   Constructive interference is the combination of two waves such that the result is of greater intensity than either wave alone, and destructive interference results in the total wave having less intensity than either wave alone.   (frame 5)

4.   Two waves must be produced at nearly the same frequency.   (frame 7)

5.   four cycles per second   (frame 8)

6.   Pitch is the physiological impression while frequency is a physical characteristic of the sound wave.   (frame 9)

7.   the number and relative intensities of overtones   (frame 11)

8.   They differ in quality.   (frame 10)

9.   It would have nodes at the ends, with maximum vibration at the center. (frame 11)

10.   Three. This is the second overtone.   (frames 11–12)

11.   a passing race car, police car, train, etc.   (frame 18)

12.   The source of sound must be moving faster than the speed of sound.   (frame 20)

13. $1\frac{1}{2}$ (frames 13–14)

14.   It is a sonic boom.   (frame 19)

PART FOUR

The Physics of Electricity and Magnetism

In our study of electricty and magnetism, we start by considering the nature of electrical charge and how it exhibits the familiar phenomena of static electricity. When electrical charges are put into motion, we have an electric current, the subject of the second chapter of this section.

At first glance, it seems that we then change topics, for we move on to magnetism. But the relationship between magnetism and electricity is extremely close, and the subject matter for the final four chapters of Part IV follows from this relationship. The last chapter, on electromagnetic waves, will lead us directly into a study of light, for light is itself an electromagnetic wave.

CHAPTER THIRTEEN

Static Electricity

Prerequisite: Chapter 5

The phenomenon of static electricity is a familiar one. It is static electricity which causes the shock when we slide across a car seat and touch the door handle, and it is static electricity which makes a sweater cling when it is taken off.

A study of this chapter will enable you to:

- compare the electrostatic forces between two charges when the distance between them is changed;
- determine the relative strengths of electric fields using Coulomb's law;
- specify the units of electric charge and electric field strength;
- interpret a diagram of electric force lines in terms of direction and relative strength;
- give examples of charging by friction and charging by induction;
- differentiate between electron action in conductors and insulators;
- describe the effect of a charged object on an electroscope;
- describe what happens to the atomic charges of the objects when two objects are charged by friction;
- calculate the force between two charges, given the strength of the charges and the distance between them (optional);
- calculate the strength of an electric field at a given distance from a given charge (optional);
- determine the electrostatic force on a given charge in a given electric field (optional).

COULOMB'S LAW

1.   The entire phenomenon of electricity is caused by the action of the charged particles which make up atoms: protons (which are positive) and electrons (which are negative).
The basic force law of electrostatics is that opposite charges attract and similar charges
repel. A negatively charged object attracts a positively charged object. Two positive objects or two negative objects, on the other hand, repel one another. Consider an atom that
has lost one electron.

(a)   Will it be positively or negatively charged? _____

(b)   Will it attract or repel an atom that has gained an extra electron?_____

— — — — — — — — — — — — — — — —.

(a) positively (it has lost a negative charge); (b) attract

2.   The strength of the attraction or repulsion force is determined by Coulomb's law:

$$F \sim \frac{Q_1 \cdot Q_2}{d^2} *$$

where Q_1 and Q_2 are the strengths of the two charges and d is the distance between them.
    This equation is very similar to the equation governing gravitational forces (Chapter
4, frame 1). Both force laws state that the force is inversely proportional to the square of
the distance between the objects.

(a)   If two charges are initially 1 meter apart and are then separated to a distance of 2

    meters, the force is reduced to what fraction of its original value? _____

(b)   If the distance is made 3 meters, the force is what fraction of the original?_____

— — — — — — — — — — — — — — —

(a) $\frac{1}{4}$; (b) $\frac{1}{9}$

3.   In Chapter 1 it was stated that all quantities could be measured in terms of distance,
time, and electrical charge. Electric charge is measured in *coulombs*, a unit independent
of the units of mass, distance, and time. A coulomb is essentially a certain number of electrons (for negative charge) or protons (for positive charge). A coulomb is the amount of
charge on $6.2425 \cdot 10^{18}$ electrons, or, in conventional notation, 6,242,500,000,000,000,000
electrons. That is a lot of electrons! (Don't memorize the number, by the way.)

(a)   What element(s) in the expression $F \sim \dfrac{Q_1 \cdot Q_2}{d^2}$ would be given in coulombs?

    _____

(b)   What element(s) would be given in meters? _____

— — — — — — — — — — — — — —

(a) Q_1 and Q_2 ; (b) d

_____

*The symbol "~" means "is proportional to."

Optional

4.   The SI unit of force is the newton, the unit of charge is the coulomb, and the unit of distance is the meter. If one coulomb of charge is located one meter away from another coulomb of charge, the electrostatic force between the two is tremendous: $9 \cdot 10^9$ newtons. This, then, is the constant which must be used to change the proportion to an equation. Thus Coulomb's law becomes:

$$F = 9 \cdot 10^9 \cdot \frac{Q_1 \cdot Q_2}{d^2}$$

What are the SI units in this equation?

F: _____ ; Q_1 and Q_2 : _____ ; d : _____

- - - - - - - - - - - - - - - -

newtons; coulombs; meters

5.   A force of 9 billion newtons is not seen in normal cases of electrostatic force, since a coulomb is a far larger charge than is generally encountered. A normal charge may be a few microcoulombs, where:

$$10^6 \text{ microcoulombs} = 1 \text{ coulomb.}$$

Suppose you charge two small combs with 0.2 microcoulombs each by combing your hair.

(a)   How much charge, in coulombs, is on each comb?_____

(b)   What force do the combs exert on each other when they are 10 centimeters

apart? _____

- - - - - - - - - - - - - - - -

(a) $2 \cdot 10^{-7}$ coulombs

(b)   0.036 newtons. $F = 9 \cdot 10^9 \dfrac{(2 \cdot 10^{-7})(2 \cdot 10^{-7})}{(0.1)^2}$

$$= 36 \cdot 10^{-3}$$

$$= 0.036$$

(Since a newton is about $\frac{1}{4}$ pound, this comes to about 0.01 pound, or somewhat more than 0.1 ounce.)

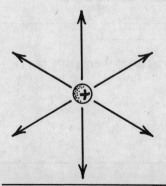

THE ELECTRIC FIELD

6.   Suppose we have a small charged object at a particular location. If another charged object were brought near that object, a force could be exerted on it, as described in Coulomb's law. It is convenient to say that the original charged object is surrounded by an "electric field." The figure at the left shows an isolated positive charge. The arrows radiating outward

from the charge indicate the direction of the force that will be experienced by a positively charged object that is brought into that area. These are sometimes called "lines of force."

(a)   Does the concentration of lines of force get more or less as they extend outward?

_____

(b)   Does the force on a positively charged object become greater or less as the object

gets farther from the charge in the figure? _____

_ _ _ _ _ _ _ _ _ _ _ _ _ _

(a) less; (b) less

7.   The figure here shows an isolated negative charge. Here the lines of force point toward the charge. Again, they show the force a positive charge would experience in that area.

(a)   What does this figure show about unlike charges?

_____

(b)   Where would the force on a positive charge be

greatest? _____

_ _ _ _ _ _ _ _ _ _ _ _ _ _

(a) they attract; (b) near the object

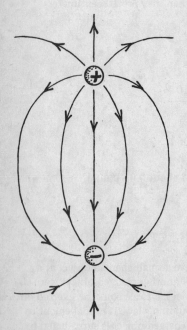

8.   As we have seen, lines of force are closer together where the force is greater. At greater distances from a charge, the lines separate—and the force diminishes. The figure in this frame shows two charges, one positive and one negative, and the force field surrounding them.

(a)   Where is the origin of each force line?

_____

(b)   Where is the destination of each force line?

_____

(c)   Where does the force seem strongest?

_____

_ _ _ _ _ _ _ _ _ _ _ _ _ _

(a) at the positive charge; (b) at the negative charge; (c) near each charge

Optional

9.   The strength of an electric field at a certain point is equal to the strength of the force on a charge of 1 coulomb placed at that point. To calculate the electric field near a charge, you assume that a charge of 1 coulomb is located at the given distance from the given charge, and you use Coulomb's law to calculate the force on this assumed 1 coulomb charge. Now use the formula $F = 9 \cdot 10^9 \, (Q_1 \cdot Q_2)/d^2$ to find the strength of an electric field at a distance of 0.5 meters from a positive charge of 3

microcoulombs. _____

— — — — — — — — — — — — — — — —

$1.08 \cdot 10^5$ newtons. $F = 9 \cdot 10^9 \; \dfrac{(3 \cdot 10^{-6})(1)}{(.5)^2}$

(This means that there is 108,000 newtons of force on each coulomb at that point in the field. We say that the field strength is 108,000 newtons/coulomb.)

10.   Once we know the strength of an electric field in newtons/coulomb, we no longer need to know the configuration of charges that caused that field in order to know what force the field will exert on a certain charge. For example, suppose we bring a charge of 0.05 coulombs to a point in an electric field where the strength is 250 newtons/coulomb. How do we find the electrostatic force experienced by the 0.05 coulomb charge? From the fact that the strength of the field at the point is 250 newtons/coulomb, we know that a force of 250 newtons will be exerted on each charge of one coulomb brought to that point. The force on our 0.05 coulomb charge is therefore 0.05 times 250, or 12.5 newtons. Notice that we did not need to know where the charge (or charges) were located which causes the electric field.

What electrostatic force would be experienced by a charge of 0.2 coulombs

placed in an electric field of 300 newtons/coulomb? _____

— — — — — — — — — — — — — — — —

60 newtons

CHARGING BY FRICTION: THE TRIBOELECTRIC SERIES

11.   We normally see the phenomenon of static electricity when two objects are rubbed together. This method is called charging by friction even though friction actually has nothing to do with the charging process. Some materials tend to hold on to their electrons more strongly than others. When two materials that differ in this tendency are placed in intimate contact (i.e., rubbed together), one gains electrons and the other loses electrons.

The table at the top of page 148 lists some materials in the triboelectric series (Note: *tribo-* means friction), which ranks materials according to their tendency to give up their electrons. Materials toward the top of the list become positively charged when placed in contact with those lower on the list.

	rabbit's fur
	glass
	wool
	cat's fur
	silk
The Triboelectric Series	felt
	cotton
	wood
	cork
	rubber
	celluloid

(a)   If rabbit's fur is used to rub a piece of rubber, how do their charges change?

_____

(b)   Would you rub felt on glass or rubber to obtain a positive charge on the felt?

_____

— — — — — — — — — — — — — — — — —

(a)   rubber gets negative and the fur gets positive
(b)   rubber (the glass would make the felt negative)

INSULATORS AND CONDUCTORS

12.   One object can cause another object to become charged by coming into physical contact with that object, when electrons move from one object to the other. The exchange of electrons, however, occurs only at the point of contact. The electrons received by an object being charged negatively may or may not distribute over the entire surface of the object, depending upon how good a *conductor* of electricity the object is. Some materials make good conductors because their electrons, rather than being bound to individual atoms, are free to move from one atom to another. The atoms in a good conductor naturally "share" their electrons so that when surplus electrons are added to one part of the material, electrons throughout the material will redistribute so that the effect is that of the charge being spread throughout the object.

A similar thing happens when a conductor is charged positively. When electrons are removed from one portion of the conductor, the remaining electrons redistribute so that the entire object is again charged.

Materials in which the electrons are not so free to move from atom to atom are *insulators*. When a part of an insulator receives excess charge, the charge remains on that part rather than distributing over the entire object. Most metals are good conductors, and most nonmetals are poor conductors—good insulators—of electricity.

Identify whether each of the following describes a conductor or an insulator.

(a)   Added electrons redistribute themselves over an object. _____

(b)   Could be used for sending electrical charges from one place to another.

_____

(c)   Any excess charge remains in position. _____

(d)   An object made up of atoms or molecules that easily share electrons.

_____

_____

(a) conductor; (b) conductor; (c) insulator; (d) conductor

THE ELECTROSCOPE

13.   The figure at the left illustrates an *electroscope*, a device that is used to detect an electric charge. The heart of the electroscope is the metal strip down the center and the flexible metal foil attached to it. This part of the electroscope is electrically insulated from the cylindrical cover with its glass end windows. The cover serves only to keep out air currents which would disturb the delicate foil. Let's use our knowledge of charge to find out what happens when you touch a charged object to the ball at the top. The ball, metal strip, and foil are good conductors.

(a)   What will happen to any charges transferred to the ball?_____

_____

(b)   Will the charges on the strip and foil be the same or different?_____

(c)   What will happen to the foil? (Hint: It is very light.) _____

_____

_____

(a)   They get distributed over the ball, strip, and foil.
(b)   same (both positive or both negative)
(c)   It will be repelled, or move away from the metal strip.

14.   As you see, the electroscope detects the electric charge on an object that touches the ball. It also, however, removes that charge from the very object it is being used to examine. There is a way around this problem.
    Suppose a negatively charged object is brought near—but not touching—the ball at the top of an uncharged electroscope.

(a)   What type of force will electrons in the ball feel, attraction or repulsion?

_____

(b)   What will these electrons try to do?_____

(c)   Where can they go? _____

(d)   What will be the resulting effect on the foil? _____

(e)   If you used a positively charged object near the ball, what type charge would get on the foil? _____

— — — — — — — — — — — — — — —

(a) repulsion; (b) leave, move away; (c) down the strip; (d) move away; (e) positive (as electrons move *up* the strip)

CHARGING BY INDUCTION

15.   We discussed earlier charging by contact with a charged object; now we will consider charging by induction in which an object is charged opposite to the object causing the charge. Suppose that, while a negatively charged object is being held near the top of an electroscope, someone touches the ball. In this case some of those electrons that are trying to escape the vicinity of the charged object will be able to flow into the hand and down into the person's body. (The body is a fairly good conductor.) Now the hand is removed. This leaves a net positive charge on the electroscope. Notice that by using a negatively charged object here we put positive charge on an electroscope, causing some of the electrons to move onto another conductor in contact with the electroscope.

Can you describe how a positively charged object can be made to cause a negative charge on an electroscope—or any other conductor? _____

_____

_____

— — — — — — — — — — — — — — —

The positively charged object is brought near the electroscope. Then when a person touches the ball, the electrons are pulled from the person's body, leaving excess electrons on the ball when the hand is removed.

SELF-TEST

The questions below will test your understanding of this chapter. Use a separate sheet of paper for your diagrams or calculations. Compare your answers with the answers provided following the test.

1.   What is the SI unit of electric charge? _____

2.   What direction does the electric field have near a negative charge? _____

_____

3.   What do the arrows represent in a drawing of an electric field? _____

_____

4.   What is shown by the relative spacing of the lines in different parts of an electric field drawing? _____

5.   What change takes place to make one object positive and the other negative when two objects are rubbed together? _____

6.   Is it possible to make one object negative without making another object less negative? Why or why not? _____
     _____

7.   What atomic particle moves when electricity moves through a conductor?
     _____

8.   What happens to the electrons in a conductor after one part of the conductor is touched by a positively charged object? _____
     _____

9.   How can an electroscope be charged negatively by induction? _____
     _____

10. Suppose two charges are located 10 cm apart and feel an attractive force of 12 units. What will be the force between the two when they are placed 5 cm apart?
     _____

11. *Optional.* Suppose two charges, each 4 microcoulombs, are located 20 cm apart. What force is exerted on each charge? _____

12. *Optional.* How much charge must be on one object for it to be repelled with a force of 5 newtons by a second charge of one microcoulomb at a distance of one meter?
     _____

13. *Optional.* Suppose the electric field at a particular location is 230 N/C. How much force will be exerted on a 3 coulomb charge at that point?_____

14. *Optional.* What is the electric field at a distance of 20 cm from a 4 microcoulomb charge? _____

15. *Optional.* What will be the force on another 4 microcoulomb charge at the location for which you calculated the electric field in question 14? Compare this answer to that of problem 11. _____

Answers to Self-Test

If your answers do not agree with those given below, review the frames indicated in parentheses before you go on to the next chapter.

1.   coulomb   (frame 3)

2.   It points radially inward toward the charge.   (frame 7)

3.   They show the direction of the force on a positive charge at that point.   (frame 6)

4.   Where the lines are closer together, the field is stronger. There would be more force on a charge there.   (frame 8)

5.   Electrons leave the object that becomes positive, and go onto the object that becomes negative.   (frame 11)

6.   No. An object is made negative by adding electrons to it. The electrons must be taken from some other object, which then becomes less negative (or more positive). (frame 11)

7.   electrons (Protons are bound in place in the nuclei of atoms.)   (frame 12)

8.   Electrons on the part of the conductor that is touched will move onto the positive object. Then electrons throughout the conductor will redistribute so that the lack of negative charge on the conductor is spread throughout. Thus, the positive charge is spread throughout.   (frame 12)

9.   A positively charged object is brought near the ball. Then the ball is touched by an uncharged conductor (such as your finger). This causes electrons to be pulled from the conductor onto the ball. Then the conductor is pulled away and the positive object is removed. The electroscope is negative.   (frame 15)

10.   48 units of force. (The distance between them is $\frac{1}{2}$ as much. The inverse square of $\frac{1}{2}$ is 4, so the force becomes 4 times as great.)   (frame 2)

11.   3.6 newtons. Solution: $F = 9 \cdot 10^9 \dfrac{Q_1 \cdot Q_2}{d^2}$

$$= 9 \cdot 10^9 \frac{(4 \cdot 10^{-6})(4 \cdot 10^{-6})}{(0.2)^2}$$

$$= 3600 \cdot 10^{-3} \text{ newtons}   \text{(frames 4-5)}$$

12. $5.55 \cdot 10^{-4}$ coulombs or 555 microcoulombs. Solution: Letting Q_2 be the unknown charge:

$$5 = 9 \cdot 10^9 \frac{(10^{-6}) \cdot Q_2}{(1)^2}$$

$$5 = 9 \cdot 10^3 \cdot Q_2$$

$$Q_2 = \frac{5}{9 \cdot 10^3}$$

$$= 0.555 \cdot 10^{-3} \text{ coulombs}   \text{(frame 5)}$$

13.   690 newtons. Solution: Multiply the electric field strength (230 N/C) by the charge located in the field (3 C).   (frame 10)

14. $9 \cdot 10^5$ N/C   Solution: Letting Q_2 be the 1 coulomb charge put into the field:

$$F = 9 \cdot 10^9 \ \frac{(4 \cdot 10^{-6})(1)}{(0.2)^2}$$

$$= 900 \cdot 10^3 \ \text{newtons}$$

$$= 9 \cdot 10^5 \ \text{newtons on one coulomb} \quad \text{(frame 10)}$$

15.   3.6 newtons. Solution: Multiply the electric field by the charge. This is the same answer that we obtained in problem 11 where we directly calculated the force on a 4 microcoulomb charge at a distance of 20 cm from another charge of 4 microcoulombs.   (frame 10)

CHAPTER FOURTEEN

Current Electricity

Prerequisite: Chapter 2, frames 1–5 for frames 6, 20, and 21

Electrons in a charged object carry the charge. When electrons move around in a conductor, we say that there is a current in the conductor. In this chapter we will see in more detail what happens when current flows. Most of the examples will use a metal wire as a conductor.

After studying the current chapter, you will be able to:

- specify the direction of electron flow from a charged object to ground;

- differentiate between electron flow and conventional current flow;

- calculate the energy released when a given amount of charge flows across a given potential difference;

- specify the unit of current and relate it to charge and time;

- recognize schematics for batteries and resistors;

- use Ohm's law to solve simple current problems;

- determine total voltage across resistors connected in series and in parallel;

- determine total resistance when resistors are connected in series and in parallel;

- specify the effect of a burn-out (open circuit) in a series or parallel circuit;

- differentiate between alternating and direct current;

- determine the energy dissipated by a circuit, given the power and time;

- given two of the three quantities—current, voltage, and resistance—calculate the power dissipated in a circuit;

- given the cost of electrical energy, calculate the cost of operating a given device for a certain amount of time;

- calculate the number of joules in a kilowatt-hour (optional).

FLOW OF ELECTRIC CHARGE

1.   Suppose one end of a wire is connected to a very large conductor, one large enough so that we can add electrons to it or take them away without appreciably changing the density of charge on the object. (This occurs because the object is large enough that when electrons are added to it, they distribute themselves over it so that relatively few are left in any one area.) Such an object is usually called "ground"; it may actually be the earth (which is, naturally, the best "ground" available). When we touch a negatively charged object to the other end of the wire, electrons move down the wire from the charged object to ground, creating an electric current. This current is very short in duration, lasting but a fraction of a second. Later we will consider methods by which a source of electrons is able to replenish its supply of electrons as they are removed, to produce a steady current in the wire. Examine the diagram on the left.

(a)   What happens to electrons when the object touches the wire? _____

_____

(b)   Why is the earth the "best ground available"?

_____

_ _ _ _ _ _ _ _ _ _ _ _ _ _ _ _

(a) They move from ground, up the wire, and into the object to balance the positive charge there. (b) It is the largest object around!

2.   Physicists speak of the direction of "conventional" current in a conductor as the direction of the flow of *positive* charges. In other words, they say that electric current flows from the positive end of a wire to the negative end. This may create a momentary confusion. But no one is claiming that protons move through the wire. We know that the electrons move.

Because it is more convenient to speak of current flow from positive to negative, and because the names "negative" and "positive" were assigned before it was known that the electrons are the current carriers, we still speak in physics of current moving from positive to negative.

Refer back to the figure in frame 1.

(a)   Which way did you say electrons move? _____

(b)   Which way does conventional current flow? _____

_ _ _ _ _ _ _ _ _ _ _ _ _ _ _

(a)   up the wire (negative to positive)
(b)   down the wire (positive to negative)

3.   We have seen that the flow of positive charges is called electric current. The unit of current is the ampere. One ampere flows past a point in a conductor when one coulomb of charge moves by that point each second. Thus:

1 ampere = 1 coulomb/second.

(a)   How many amperes of current are caused by 2 coulombs of charge flowing in 0.1

second?_____ $2/.1^5 = 20$ amp _____

(b)   How many amperes result from 60 microcoulombs flowing in 2 seconds? (1 micro-

coulomb = 10^{-6} coulombs) _____ $\dfrac{60 \cdot 10^{-6}}{2}$ _____ $= 30 \cdot 10^{-6} = 3 \cdot 10^{-5}$ _____

(a) 20 amperes; (b) $3 \cdot 10^{-5}$ amps, .00003 amps, or 30 microcoulombs (See Appendix I
for a review of powers of 10.)

BATTERIES AND VOLTAGE

4.   Batteries are a familiar source of electric charge that causes current to flow in wires.
Chemical reactions that take place in the batteries separate negative and positive charges,
causing one terminal of the battery to be positive and the other negative. When the two
terminals of the battery are connected (by a wire), electrons flow. If this flow is not too
great, chemical action in the battery continues to take electrons from one terminal and
place them on the other, maintaining the original charge.

(a)   From which terminal are electrons taken by the chemical action of the battery?

_____

(b)   From which terminal do electrons leave to flow through the wire?_____

(c)   In what direction does the conventional current flow through the wire? _____

_____

(a) positive; (b) negative; (c) positive to negative

5.   You may have noticed that batteries are marked in "volts." The common flashlight
battery, for example, is $1\frac{1}{2}$ volts. The voltage of a battery tells us how much more energy
each coulomb of charge has at one terminal than it has at the other. In moving from one
terminal to the other, the charge loses this energy, which may be converted to heat and
light. Thus when one coulomb of positive charge flows from the positive to the negative
terminal of a $1\frac{1}{2}$ volt battery, it gives up a certain amount of energy. If *two* coulombs
flow across the $1\frac{1}{2}$ volt battery, twice as much energy is lost. *Three* coulombs—three
times the energy given up. The voltage is equal to the energy given up by each coulomb
of charge; it is the energy per coulomb.

(a)   If one coulomb of charge flows across a 3 volt battery, how does the energy given
up compare to that given up by one coulomb moving across a $1\frac{1}{2}$ volt battery?

_____

(b)   Does an electron have more energy at the positive or negative terminal? _____

(c)   Does positive charge have more energy at the positive or negative terminal?

_____

(a) twice as much; (b) negative, since it gives up energy as it flows to the positive terminal; (c) positive

6.   The term "voltage" may describe energy per charge, but the more technical term is *potential difference*. The potential difference across the terminals of the standard flash-light battery is $1\frac{1}{2}$ volts.

The amount of energy a coulomb of charge loses when it falls across a potential difference of one volt is one joule. (Since positive current naturally goes from positive to negative like an object naturally falls from above to below, we often speak of charge "falling.")

$$1 \text{ volt} = 1 \text{ joule/coulomb}$$

(a)   Three coulombs of charge flow across a $1\frac{1}{2}$ volt battery. How much energy is lost by the electric charge? _____

(b)   How much charge must flow between the terminals of a 3 volt battery to release 12 joules of energy? _____

– – – – – – – – – – – – – –

(a) 4.5 joules (3 · 1.5); (b) 4 coulombs

OHM'S LAW

7.   Normally, one does not connect a wire straight from one terminal of a battery to the other, because this would cause too much current to flow, and the battery could not restore the charge to its terminals at the rate necessary. The battery would "run down." Besides, the circuit would accomplish nothing useful. Suppose, instead, that a light bulb is placed in the circuit as shown in Figure 14.1 (part 1).

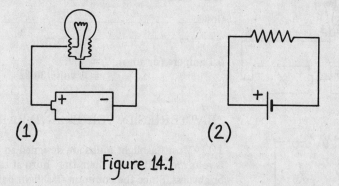

(1)   (2)

Figure 14.1

Sequence the items below in the direction in which conventional current will flow.

_5_ negative terminal   _4_ wire to right of lamp

_1_ positive terminal   _3_ lamp

_2_ wire to left of lamp

– – – – – – – – – – – – – –

Current will flow from positive terminal, to wire to left of lamp, to lamp, to wire to right of lamp, to wire to right of lamp, to negative terminal.

8.   We normally use symbols to represent the various circuit components. Figure 14.1 (part 2) shows the circuit schematically. The sawtoothed line (∧∧∧) is used to represent the light bulb in this case, but it may represent any device which offers resistance to the flow of electric current and limits the amount of current which will flow when a certain voltage is applied. We call this component a "resistor."

What effect would you expect a resistor of greater resistance to have on current

flow? _____

– – – – – – – – – – – – – – – –

cut it down, reduce it

9.   The relationship between voltage, current, and resistance is given by Ohm's law:

Voltage (in volts) = current (in amperes) · resistance (in ohms)

or symbolically:

$$V = IR$$

where I represents electric current and resistance R is measured in ohms (pronouned to rhyme with "homes").

Although voltage is defined as being the energy per coulomb, it can be thought of loosely as being the "pressure" which causes the electric current to flow. Just as more water pressure causes an increase in the amount of water which flows through pipes, more voltage causes more current to flow. But when resistance is increased, less current flows.

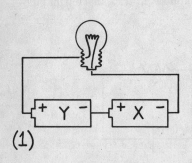

(1)

Suppose you use a 3 volt battery to light a lamp which has a resistance of 30 ohms (symbolically: 30 Ω). Use $V = IR$ to determine how much current

flows. _____

– – – – – – – – – – – – – – – –

0.1 ampere (or amp). $I = V/R$
                                   $= 3$ volts$/30\ \Omega$

BATTERIES IN SERIES AND PARALLEL

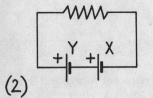

(2)

10.   Most flashlight bulbs are designed so that when 3 volts are used across them they burn at the correct brightness. Since the common flashlight battery is $1\frac{1}{2}$ volts, two batteries are connected together as shown in part 1 of the figure here. Part 2 shows this in schematic form. (Notice the symbol for a battery.) The positive terminal of battery X is $1\frac{1}{2}$ volts higher in

voltage than the negative terminal, and the negative terminal of battery Y is connected directly to this. Then the positive terminal of this battery is another $1\frac{1}{2}$ volts higher in voltage. These batteries are connected positive-to-negative, and they are said to be connected "in series."

(a)   What is the total voltage (potential difference) across these two batteries in series?

_____

(b)   In general, how would you find the overall voltage of batteries connected in series?

_____

_ _ _ _ _ _ _ _ _ _ _ _ _ _ _ _

(a) 3 V; (b) add the individual voltages

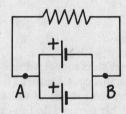

11.   The figure in this frame shows two batteries hooked up side-by-side (positive-to-positive and negative-to-negative). Batteries connected like this are said to be "in parallel." The voltage produced this way is the same as the voltage of one of the batteries. The advantage of such a hook-up is that the batteries will last longer or will be able to supply more current without running down.

(a)   How does the effective voltage of two batteries connected in series compare to that of the same two batteries connected in parallel? _____

(b)   If the batteries of the figure are each $1\frac{1}{2}$ volt, what is the potential difference between points A and B?_____

_ _ _ _ _ _ _ _ _ _ _ _ _ _ _ _

(a) double; (b) $1\frac{1}{2}$ volts

RESISTORS IN SERIES AND PARALLEL

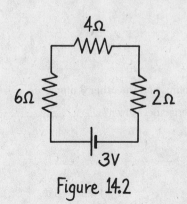

Figure 14.2

12.   Figure 14.2 shows three resistors connected in series. In a series connection, current must go through all three of the resistors to get from one side of the battery to the other. The total resistance of a series circuit is the sum total of the resistances of each of the resistors.

(a)   What is the total resistance of the circuit of Figure 14.2? _____

(b)   Use Ohm's law to find how much current will flow. _____

_ _ _ _ _ _ _ _ _ _ _ _ _ _ _ _

(a) 12 ohms; (b) 0.25 amps $(I = V/R = 3/12 = 0.25)$

13.   The total voltage across a number of resistors in series is equal to the sum of the voltage drops across the individual resistors. Let's calculate the voltage (potential difference) across just one of the resistors of Figure 14.2. Consider the 2 ohm resistor. The current through it is 0.25 amps. (All of the current flows through the entire circuit.) The voltage necessary to force 0.25 amps through 2 ohms is found using Ohm's law again:

$$V = 0.25 \text{ amps} \cdot 2 \text{ ohms}$$
$$= 0.50 \text{ volts}$$

(a)   What is the voltage across the 4 ohm resistor? _____

(b)   What is the voltage across the 6 ohm resistor? _____

(c)   What is the sum of the voltages across all three resistors? _____

(d)   How does this compare to the voltage across the battery? _____

– – – – – – – – – – – – – – – –

(a) 1 volt; (b) 1.5 volts; (c) 3 volts; (d) the same

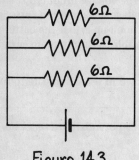

Figure 14.3

14.   Figure 14.3 shows resistors connected side-by-side, "in parallel." In a parallel circuit, current leaving one terminal of the battery has a choice of paths which it can take to get to the other terminal. In this case, each resistor has a resistance of 6 ohms. Since the current has three paths to take, it divides. The effective resistance of resistors in parallel is always less than the resistance of any one of the resistors. The effective resistance R in a parallel circuit is given by the equation:

$$\frac{1}{R} = \frac{1}{R_1} + \frac{1}{R_2} + \frac{1}{R_3} + \cdots$$

where R_1, R_2, and so on are the resistances of the individual resistors.

   Calculate the effective resistance of the three resistors in the figure in this frame.

_____

– – – – – – – – – – – – – – – –

2 ohms $\left(\dfrac{1}{R} = \dfrac{1}{6} + \dfrac{1}{6} + \dfrac{1}{6} = \dfrac{3}{6}; \; R = \dfrac{6}{3} = 2 \right)$

15.   Suppose you have two resistors in parallel, one 2 ohms and the other 3 ohms.

(a)   How will the effective resistance compare to the resistances given? _____

_____

(b)   What is the effective resistance? _____

– – – – – – – – – – – – – – – –

(a) less than either; (b) $\frac{6}{5}$ ohms

16.   Follow the steps below to solve the problem represented by the circuit at the left.

(a)   Are the resistors in series or in parallel?

_____

(b)   Are the batteries in series or in parallel?

_____

(c)   What is the potential difference across the battery combination? _____

(d)   What is the resistance of the circuit? _____

(e)   What current flows from the batteries? _____

(f)   What current flows through the 12 ohm resistor? _____

(g)   What current flows through the 4 ohm resistor? _____

(h)   Is the answer to question (e) equal to the total of the answers to (f) and (g)? Why?

_____

_____

– – – – – – – – – – – – – – –

(a) parallel; (b) series; (c) 6 volts; (d) 3 ohms; (e) 2 amps (6 volts/3 ohms); (f) 0.5 amp (6 volt/12 ohms); (g) 1.5 amps; (h) It should be, because the current divides to go through the two resistors, and the sum of the parts equals the whole.

SERIES VS. PARALLEL: HOME WIRING

17.   Let's look at what happens when a light bulb burns out. A lamp burns out when the tiny hot wire inside it—the filament—breaks. (The molecules of the filament gradually evaporate directly from solid to gas as the lamp burns. Thus the filament gets thinner and thinner until it breaks.) Now look at Figure 14.2 and imagine the wire within one of the lamps (the resistors) breaking.

(a)   What happens to the other two lamps? _____

(b)   Does it make any difference which light bulb burns out? _____

(c)   Suppose all your Christmas tree lights go out when a single bulb is removed. Are

these wired in series or in parallel? _____

– – – – – – – – – – – – – – –

(a) they go out; (b) no (they all go off no matter which burns out); (c) series

18.   Now consider the circuit of Figure 14.3. When one of the bulbs burns out in this circuit, current can still flow through the other branches. The fact that current no longer flows through one branch of a parallel circuit does not prevent it from getting from one terminal of the battery to the other by means of the other circuits.

(a)   If two bulbs burned out in this circuit, what would happen to the third?

(b)   Is the current in your home wired in series or in parallel?_____

——————————————————

(a) It would continue to burn; (b) parallel (When one electrical device burns out—or is turned off—the rest still work.)

ALTERNATING CURRENT

19.   We have been discussing direct current (DC), the kind supplied by a battery. The current we use every day in our homes is alternating current (AC). Both types of current consist of electrons moving through the wire, but in house wiring the electrons move first one direction, then the other. In order for this to happen the ends of the wires supplying the current alternately change back and forth between positive and negative.

The alternating current system used in the United States changes from positive to negative and back to positive 60 times each second. Its frequency is therefore 60 cycles per second, or 60 hertz. In Europe, 50 cycle/second alternating current is used.

As far as light bulbs are concerned, the difference between direct and alternating current is unimportant, for it is the flow of current which makes the filament heat up. It does not matter to the bulb whether or not the current changes direction. Devices which contain motors, or electronic devices such as radios, stereos, and TVs, are made to operate on either AC systems or on DC systems. If they are designed to be used on both, they must contain a circuit which changes one type current to the other. (Normally the AC is changed into DC when portable radios and tape players are plugged into a socket in your home.)

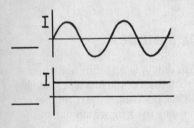

(a)   What type of current is used in a portable calculator?_____

(b)   What type of current is used to run an electric table saw?_____

(c)   Label each of the graphs here as representing AC or DC.

——————————————————

(a) direct; (b) alternating; (c) top—AC, bottom—DC

ENERGY DISSIPATED IN A RESISTOR

20.   We said earlier that voltage is the energy per coulomb. As each coulomb flows through a resistor, this energy shows up in the resistor as heat (and light, if the resistor gets hot enough—as in the light bulb). The rate at which energy is transferred to heat and light is called power, and the unit of power is the watt.

power = energy ÷ time

A 40 watt light bulb will use less energy than a 60 watt bulb in the same amount of time. To calculate the amount of energy used, we multiply power by the time during which the power is used. If power is expressed in kilowatts (one kilowatt = 1000 watts) and time in hours, then the unit of energy is kilowatt-hour (abbreviated kWh).

(a)   How many kilowatt-hours are used in burning a 100 watt bulb for 8 hours? _____

(b)   Since the unit of kilowatt-hours appears on an electric bill, are you paying for

power or energy? _____

(c)   Suppose you pay 15¢ per kilowatt-hour. How much does it cost to watch a 300

watt color TV for 6 hours? _____ $P = \dfrac{E}{T}$

- - - - - - - - - - - - - - - -

(a) 0.8; (b) energy; (c) $.27 (or 27¢) (300 W = .3 kW; 0.3 kW · 6 hrs = 1.8 kWh;
1.8 kWh · 15¢/kWh = 27¢)

SI CALCULATIONS

21.   As you recall, one volt = 1 joule/coulomb and one ampere = 1 coulomb/second.
When we multiply potential difference (volts) by current (amperes), we obtain:

$$\frac{\text{joules}}{\text{coulomb}} \cdot \frac{\text{coulomb}}{\text{second}} = \text{joules/second.}$$

One joule/second = one watt. The product of the potential difference across a resistor
and the current through the resistor is the power input into the resistor. In equation form:

$$P = IV.$$

Ohm's law is $V = IR$. Thus, when we substitute this into the power equation we obtain
$P = I^2R$.
    From Ohm's law, $I = V/R$. Substituting V/R for I in the first equation, we obtain
$P = V^2/R$.
    These three equations for power are equivalent, but sometimes one is more conve-
nient to use than another.

(a)   To calculate the resistance of a light bulb which, when used in a 115 volt circuit,

uses 100 watts of power, which formula will you use? _____

(b)   What is the answer? _____

(c)   What current flows through the lamp? _____

- - - - - - - - - - - - - - - -

(a) $P = V^2/R$; (b) 132 ohms; (c) 0.81 amps (From $P = IV$ or $V = IR$ or $P = I^2R$. The
first is preferred because it does not rely on your calculation of R.)

Optional

22.   Work through the steps below to discover the number of joules in a kilowatt-
hour.

(a)   How many watt-hours are in a kilowatt-hour? _____

(b)   How many seconds are in an hour? _____

(c)   If a watt is a joule/second, what is a joule (in terms of watts and seconds)?

_____

(d)   One kilowatt-hour is how many joules? _____

– – – – – – – – – – – – – – – –

(a) 1000; (b) 3600; (c) one watt-second; (d) 3.6 · 10^6, or 3,600,000

SELF-TEST

The questions below will test your understanding of this chapter. Use a separate sheet of paper for your diagrams or calculations. Compare your answer with the answers provided following the test.

1.   What does the voltage of a battery tell us about the electric charge on the terminals

of the battery? _____

_____

2.   Suppose a 9 volt battery is connected across a 100 ohm resistor. What current will

flow through the resistor? _____

3.   If a greater voltage is applied across a resistor, (more/the same/less) _____
current will flow.

4.   Are the batteries in a conventional flashlight connected in series or in parallel?

_____

5.   A "power beam" flashlight contains 5 regular D-cell batteries ($1\frac{1}{2}$ volts each) in

series. What is the voltage across these 5 batteries? _____

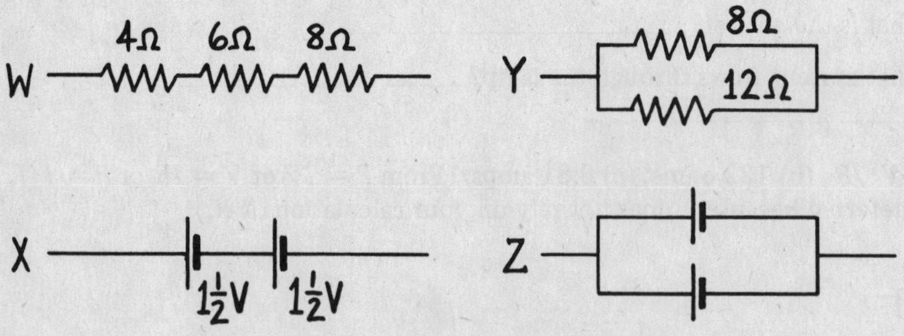

Figure 14.4

6.   (a)   Which of the parts of Figure 14.4 represents resistors in series? _____
      (b)   Which represents batteries in series? _____
      (c)   Which represents resistors in parallel? _____

7. (a)   What is the effective resistance of Figure 14.4, part Y? _____

    (b)   What is the effective resistance of Figure 14.4, part W? _____

8. How does the effective resistance of four resistors in parallel compare to the individual resistances? _____

9. As more and more resistors are added in parallel, what happens to the total current flowing from a source? (This occurs when you continue to plug in more and more devices in your home.) _____

10. If electrical energy costs 10¢ per kilowatt-hour, how much does it cost to operate a 1200 watt hair dryer for $\frac{1}{2}$ hour? _____

11. One half ampere flows from one terminal of a 3 volt battery to the other terminal for 5 seconds. How much energy is released? _____

12. Suppose an auto headlight is on for 1 hour. Two amperes of current flows through it from the 12 volt battery of the car.

    (a)   How many coulombs of charge flow through the headlight during the hour?

          _____

    (b)   What energy does this charge release (in joules)? _____

    (c)   What is the power in watts of the headlight used like this? _____

13. One half amp flows through a 40 ohm resistor. How much power is being used? What is the voltage across the resistor? _____

14. Suppose the battery connection of Figure 14.4, part X is connected to the resistors of Figure 14.4, part W. How much current flows? What is the voltage across the 6 ohm resistor? _____

Answers to Self-Test

If your answers do not agree with those given below, review the frames indicated in parentheses before you go on to the next chapter.

1. It tells us how much more energy the charge has on one terminal than on the other. Voltage is energy per charge.   (frame 5)

2. 0.09 ampere $(V = IR; 9 = I \cdot 100)$ (frame 9)

3. more (This follows from Ohm's law.)   (frame 9)

4. series   (frame 10)

5. 7.5 volts   (frame 10)

6. (a) W   (frame 12);   (b) X   (frame 10;   (c) Y   (frame 14)

7. (a) 4.8 ohms. $\dfrac{1}{R} = \dfrac{1}{R_1} + \dfrac{1}{R_2}$

$$= \dfrac{1}{8} + \dfrac{1}{12}$$

$$= \dfrac{3}{24} + \dfrac{2}{24}$$

$$= \dfrac{5}{24}$$

$$R = \dfrac{24}{5} = 4.8 \quad \text{(frames 14–15)}$$

(b) 18 ohms   (frame 13)

8. The effective resistance is less than the resistance of any of the resistors. (frame 14)

9. The total current increases, because the effective resistance decreases. (In your home, too much current will blow a buse.)   (frames 14–16)

10. 6¢. 1200 watts = 1.2 kilowatt
    1.2 kW · .5 hr = .6 kWh
    .6 kWh · 10¢/kWh = 6¢   (frame 20)

11. 7.5 joules ($\frac{1}{2}$ amp is $\frac{1}{2}$ coulomb each second. In 5 seconds this is 2.5 coulombs, and since the potential difference is 3 volts, or 3 joules/coulomb, this amounts to 3 · 2.5, or 7.5 joules.)   (frames 3, 6)

12. (a) 7200 coulombs (2 amps = 2 coulombs/second; 1 hr = 3600 sec) (frame 3)
    (b) 86,400 joules (7200 coulombs · 12 joules/coulomb)   (frame 6)
    (c) 24 watts ($P = IV$)   (frame 21)

13. 10 watts; 20 volts ($P = I^2R = .5^2 \cdot 40 = 10$ watts; $V = IR = .5 \cdot 40 = 20$ volts) (frame 21)

14. 0.167 amp; 1 volt ($V = IR$, $I = \dfrac{V}{R} = \dfrac{3}{18} = .167$ amp; $V = IR = .167 \cdot 6 = 1$ volt) (frames 9, 10, 12, 13)

CHAPTER FIFTEEN

Magnetism and Magnetic Effects of Currents

Prerequisite: Chapter 14

A magnet has a north and a south pole; like poles repel and unlike poles attract. This rule of attraction and repulsion is very similar to the one that applies to electric charges. (In fact, the equation for the force between two magnetic poles has the same form as the Coulomb's law equation—and the law of gravity.) The connection between magnetism and electricity is much closer than just the similarity of force laws, and most of this chapter will be spent discussing this connection.

After finishing this chapter, you will be able to:

- interpret a magnetic field drawing in terms of relative strength and directions of forces on north and south poles;

- describe the magnetic field of the earth and specify what magnetic poles are located at the earth's north and south geographic poles;

- use the dot-and-X convention to indicate directions into and out of a surface;

- indicate the direction of magnetic field produced when current passes through a wire;

- specify the direction of magnetic field due to current in a loop or solenoid;

- state the effect of an iron core in a solenoid;

- state the difference between a magnetized and an unmagnetized object in terms of domains;

- specify the basic difference in construction between meters and motors;

- state the nature of the force by which both electric meters and motors operate;

- specify, on the atomic level, what causes magnetism in magnetic materials;

- describe the method used to change current direction in the loop of an AC motor and a DC motor;

- calculate the strength of magnetic field in the case of long straight wires and in the case of solenoids (optional).

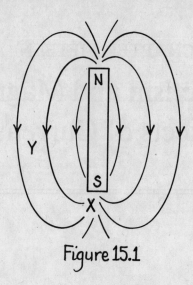

Figure 15.1

MAGNETIC FIELDS

1.   Figure 15.1 shows a bar magnet with its north and south poles marked. The lines with arrowheads represent the magnetic field of the magnet. The direction of this field at any point is defined as the direction of the force on a north pole placed at that point. As in the electric field, the relative density of the lines tells us the relative strength of the field.

(a)   At which point, X or Y, would another north pole experience the greater force?____

(b)   Would this force be one of attraction or repulsion? _____

— — — — — — — — — — — — — — —

(a) X (greater density of lines); (b) attraction (unlike poles)

2.   Suppose we have a small bar magnet supported from its center so that it is free to swing about that center. Now suppose we locate this magnet at point Y in the magnetic field of Figure 15.1. The south pole of this small suspended magnet will feel a force toward the north pole, and the north pole will feel a force in the opposite direction. The suspended magnet will therefore align itself along the magnetic field of the bar magnet. In practice, this is one method used to plot the magnetic field of a magnet. (You might do such an exercise in an elementary physics laboratory.) A more common use of such a device is as a compass.

       The figure at the top of page 169 shows that the earth has a magnetic field, in a form similar to that of a bar magnet. When a magnet pivoted at its center—a compass—is held in this field, the north pole of this magnet points in a direction which we call north. If you have ever used a magnetic compass, you know that the end of the compass which we call "north" points toward the geographic north of the earth.

       What type magnetic pole is located near the geographic north pole of the earth?

_____

— — — — — — — — — — — — — —

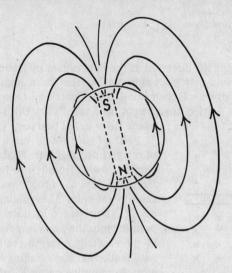

a south magnetic pole (This probably seems strange, but if we call the end of the compass that points north a north magnetic pole, no other answer is possible, for opposite poles attract.)

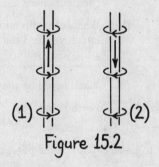

Figure 15.2

FIELDS PRODUCED BY CURRENTS

3.   A bar magnet is only one way to produce a magnetic field. Magnetic fields are produced every time an electric current flows in a wire. Figure 15.2 (part 1) shows a wire with a current flowing upward. The lines represent the magnetic field which surrounds the wire. The field's direction is around the wire, perpendicular to the direction of the current. In front of the wire (between you and the page) the direction of the field is from left to right, while on the back side of the wire the field is from right to left.

Figure 15.2 (part 2) shows a current going downward in a wire.

(a)   How does the direction of the magnetic field in part 2 compare to the direction of

the current flow? _____

(b)   How does the direction of the magnetic field in part 2 compare to that in part 1?

_____

– – – – – – – – – – – – – – –

(a) perpendicular, around the wire; (b) opposite

4.   You can determine the direction of the field caused by a current in a wire by imagin-
ing that you grab the wire with your right hand so that your thumb points in the direction
of the current. Your fingers will then curve in the direction of the magnetic field. (We'll
call this technique the "curled fingers right-hand rule.") Try this to verify the directions
of the fields in the two parts of Figure 15.2.

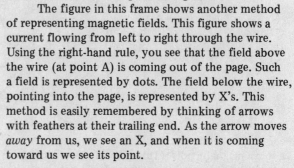

The figure in this frame shows another method
of representing magnetic fields. This figure shows a
current flowing from left to right through the wire.
Using the right-hand rule, you see that the field above
the wire (at point A) is coming out of the page. Such
a field is represented by dots. The field below the wire,
pointing into the page, is represented by X's. This
method is easily remembered by thinking of arrows
with feathers at their trailing end. As the arrow moves
away from us, we see an X, and when it is coming
toward us we see its point.

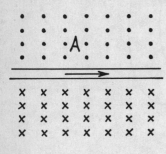

(a)   Use the right-hand rule to determine the direction of the magnetic field at Y near the wire in
the figure here. _____

(b)   What is the direction of the magnetic field between you and the wire in this case?

_____

(c)   What would be the direction of the force on a
north pole at point Z? _____

– – – – – – – – – – – – – – –

(a) down into the page; (b) left to right; (c) out of the page, toward you

Optional   FIELD CALCULATIONS

5.   The unit of magnetic field is the tesla (T).* When one ampere flows in a long
straight wire, the magnetic field one meter from the wire is $2 \cdot 10^{-7}$ tesla. In general,
the equation for the magnetic field B (in tesla) at a distance r (in meters) from a wire
carrying a current I (in amperes) is:

$$B = 2 \cdot 10^{-7} \frac{I}{r}$$

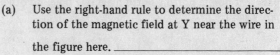
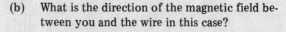

What is the magnetic field at a distance of 1 centimeter from a long straight wire
carrying a current of 8 amperes? _____

– – – – – – – – – – – – – – –

$1.6 \cdot 10^{-4}$ T (Be sure to put distance in meters.)

_____

*An older name for this unit is the weber per square meter (Wb/m²).

MAGNETIC FIELDS DUE TO LOOPS AND SOLENOIDS

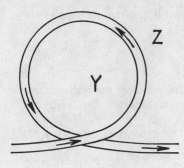

6.   Imagine grabbing the wire of the loop shown in the figure here with your right hand so that your thumb points along the current direction. You will find that no matter where you grab the wire of the loop, the magnetic field *inside the loop* points up out of the paper. Every element of the loop contributes to the field coming out of its center.

(a)   What is the direction of the field at point Y?

_____

(b)   At point Z? _____

– – – – – – – – – – – – – – – – –

(a) out of the page toward you; (b) into the page (Although the contribution at Z due to the left side of the loop is opposite to this, it is weaker because the left of the loop is farther away.)

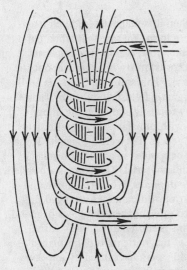

7.   The figure in this frame shows a solenoid, which can be thought of as a number of loops stretched out along the axis of the loops. Again, each element of each loop contributes to the field through the center. Because of this a solenoid can be used to produce intense fields.

Notice that the magnetic field *external* to the solenoid is similar to that of the bar magnet of Figure 15.1. If a metal which is easily magnetized and demagnetized (soft iron, for example) is placed in the loop of a solenoid, it will become magnetized when current flows through the wire of the solenoid. The additional magnetism will contribute additional magnetic field to the region surrounding the solenoid. The solenoid and enclosed iron core thus forms a very strong magnet. Such *electromagnets* are used on the end of large cranes to pick up magnetic materials.

(a)   Under what conditions would a solenoid with an iron core function as a magnet?

_____

(b)   Would a solenoid without a core function as a magnet? _____

(c)   How would you stop an electromagnet from acting as a magnet? _____

_____

– – – – – – – – – – – – – – – – –

(a) when current flows; (b) yes, but not as strong; (c) turn off the current

Optional

8.   The magnetic field at the center of a long solenoid of n turns and length ℓ is given by the equation:

$$B = 4\pi \cdot 10^{-7} \frac{nI}{\ell} *$$

You can see from the equation that the diameter of the loop does not really affect the field. (The diameter must be small compared to its length, however. This is why we specify a long solenoid.)

A typical 10 cm long solenoid might have 1000 turns of wire carrying one half amp. What is the magnetic field in the center of such a solenoid?_____

‒ ‒ ‒ ‒ ‒ ‒ ‒ ‒ ‒ ‒ ‒ ‒ ‒ ‒ ‒

$6.3 \cdot 10^{-3}$ T $\left(B = 12.6 \cdot 10^{-7} \cdot \dfrac{1000 \cdot .5}{.1} \right)$

THE CAUSE OF MAGNETISM

9.   Let's consider what is different about a material when it is magnetized. Such magnetism is caused by an electric current moving in a circle, a loop. We saw that when current moves in a circle, there is a magnetic field through the circle, making the circle act like a small magnet. In an atom, the electric charges (electrons) move in a circle about the nucleus, and in addition, each electron spins on its own axis. This spinning forms an additional magnetic field. In atoms of most materials, the magnetic fields produced by these rotating and spinning electrons cancel one another out by rotating and spinning in opposite directions, but a material that is magnetic has atoms with unbalanced magnetic fields.

A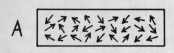

Thus the atom itself acts as if it were a magnet. These atoms tend to align themselves so as to form small magnetic areas within a material. These areas are called *domains*, or *magnetic domains*. When the north and south poles of the various domains are arranged randomly with respect to one another, the piece of metal is not magnetized. This is shown in figure A at the left. When the piece of magnetic material is placed in a strong magnetic field—perhaps caused by another magnet—these domains swing around so that they align with the north pole of one contacting the south pole of another. Figure B shows this orderly arrangement which causes the material to act as a magnet.

B

C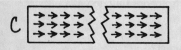

(a)   Suppose you break a permanent bar magnet into two pieces, as in figure C. Is each piece a separate pole?_____

(b)   Suppose figure A were inside a solenoid. What would happen to its domains when current was turned on? _____

‒ ‒ ‒ ‒ ‒ ‒ ‒ ‒ ‒ ‒ ‒ ‒ ‒ ‒

_____

*4π is approximately 12.6.

(a)   no, each piece has both poles (Notice that on the broken end of the left fragment, there are exposed north poles of domains, and south poles show on the broken edge of the right fragment.)

(b)   they line up, as shown in figure B

ELECTRIC METERS

10.   Consider what will happen if a current is passed through the loop of the figure here.

This loop is between north and south poles of a magnet (which could either be a horseshoe magnet or two separate magnets). When current flows as shown, a magnetic field directed out of the page will be formed by the current. So we can consider the current loop to act like a small magnet with its north pole above the plane of the paper. (Recall from frame 7 the similarity between the field of a loop or solenoid and that of a bar magnet.) Such a magnet would experience a force on its north pole toward the south pole of the surrounding magnet and a force on its south pole toward the north pole of the surrounding magnet. Thus the loop feels a twist on it, and if it is free to do so, it will turn 90° and align with the larger magnet.

In a meter, however, it is restrained by a spring from turning all the way into alignment. A pointer is connected to the loop so as to indicate the amount of twist produced by the current through the wire of the loop. Both ammeters (to measure current) and voltmeters (to measure voltage) work on this principle.

(a)   Consider the twisting loop. What force will stop it from twisting? _____

_____

(b)   What determines the amount of magnetic force twisting the loop? _____

_____

(c)   How does increased current affect the twisting force? _____

— — — — — — — — — — — — — —

(a) force exerted by the spring; (b) the strength of the large magnet and the current in the loop; (c) increases it

ELECTRIC MOTORS

11.   It is a fairly small step from a meter to a motor. In the motor there is no spring to keep the loop (or coil) of wire from rotating, so it rotates until it aligns with the large magnet. Its connection to a power supply is such that, when it achieves this alignment,

the direction of the current through the loop is reversed. This causes the loop to be forced away from that position and around $180°$ farther. At this point the current direction is again made to change so that the loop must again rotate $180°$ to come back into alignment.

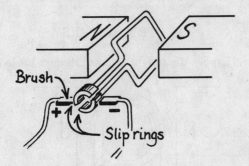

How is the current direction changed? There are two methods. The one used on motors designed for use with DC current is shown above. Each end of the wire loop is connected to a semicircular piece of metal (called a "slip ring") into which the current flows from contacts (called 'brushes") which rub it. The current enters the ring contacting the left brush (marked "+," perhaps being connected to the positive side of a battery). It then passes through the loop and emerges from the other ring into the brush touching it. Since the rings turn with the loop, first one slip ring and then the other is in contact with the left brush. Thus the current through the loop changes direction every time the loop turns through $180°$.

The current can be made to change more easily by using alternating current, which regularly changes direction. Since the speed of rotation of the loop is dependent upon the frequency of alternation of the current, such motors are used where a regulated speed is important, as in clocks, for example. AC motors which are intended to produce power rather than regulated speed are wired differently, using electromagnets controlled by the current rather than permanent magnets around the coil.

(a)   What is the basic difference between a meter and a motor? _____

_____

(b)   Is it true that a motor run by DC must have current changing direction regularly

through the loop? Why, or why not? _____

_____

— — — — — — — — — — — — — — — —

(a) in the motor, the loop spins freely and current direction changes; (b) yes, in order to have a changing magnetic field produced by the loop

SELF-TEST

The questions below will test your understanding of this chapter. Use a separate sheet of paper for your diagrams or calculations. Compare your answers with the answers provided following the test.

1. What is the general law of attraction and repulsion between magnetic poles? _____
   _____

2. What defines the direction of the magnetic field? _____
   _____

3. What does the relative density of lines of a magnetic field tell us about the strength of the field? _____

4. What type magnetic pole is found near the geographic south pole of the earth?
   _____

5. Suppose current is flowing from the bottom of this page to the top along a wire running up the right side of the page. What is the direction of the field at this point
   (·)? _____

6. The figure below represents a loop of wire carrying current in a clockwise direction. What is the direction of the magnetic field in the center of the loop? _____

7. Refer back to the figure in frame 7. If this solenoid were to be replaced by a bar magnet to cause a similar field, would the north pole of the magnet be at the top or bottom? _____

8. What two actions do electrons have which cause materials to be magnetic? _____
   _____

9. What determines how far the loop of wire in an electric meter rotates when current is sent through it? What prohibits it from rotating into complete alignment with the external field? _____
   _____

10.   What are two ways in which the current in the coils of electric motors is made to change directions (in different types of motors)? _____

_____

11.   *Optional.* What current must flow in a long straight wire in order to produce a magnetic field of 10^{-5} tesla at a distance of 2 centimeters from the wire?_____

12.   *Optional.* A 20 cm long solenoid having 150 turns has a 2 amp current in it. What is the magnetic field at its center? _____

Answers to Self-Test

If your answers do not agree with those given below, review the frames indicated in parentheses before you go on to the next chapter.

1.   Like poles repel and unlike poles attract.   (chapter introduction and frame 1)

2.   The direction of the magnetic field at any point is the direction of the force on a north pole at that point.   (frame 1)

3.   Where the lines are more closely spaced, the field is stronger.   (frame 1)

4.   north   (frame 2)

5.   out of the page toward the reader   (frame 4)

6.   into the page, away from you   (frame 4)

7.   the top   (frames 1, 7)

8.   An electron spins on its own axis, and each electron revolves around the nucleus of the atom.   (frame 9)

9.   the amount of current flowing (and the strength of the external magnet); the spring limits the rotation   (frame 10)

10.   Direct current can be changed in direction through the loop by slip rings rotating with the loop, or the motor can be designed to use the changes in current direction of alternating current.   (frame 11)

11.   1 amp. $B = 2 \cdot 10^{-7} \dfrac{I}{r}$

$$I = B \cdot \frac{r}{(2 \cdot 10^{-7})}$$

$$I = 10^{-5} \cdot \frac{.02}{(2 \cdot 10^{-7})} \quad \text{(frame 5)}$$

12. $1.89 \cdot 10^{-3}$ tesla. $B = 12.6 \cdot 10^{-7} \dfrac{nI}{\ell}$

$$= 12.6 \cdot 10^{-7} \cdot 150 \cdot \frac{2}{.2} \quad \text{(frame 8)}$$

CHAPTER SIXTEEN

Electrical Induction

Prerequisites: Chapters 14 and 15

In the last chapter we saw that when an electrical current exists in a wire, a magnetic field surrounds the wire. In this chapter we will see that a simple expansion of this idea allows us to explain the phenomenon of currents being produced by magnetic fields.

A careful study of this chapter will enable you to:

- use the right-hand rule to determine the direction of force on a current-carrying wire in a magnetic field;

- determine the direction of induced current in a wire moving in a magnetic field;

- specify the effect a changing magnetic field has on a conductor;

- use Lenz's law in determining the direction of induced current;

- describe what happens when direct current is turned on and off through one of the coils of a transformer;

- specify the reason an iron core is used in a transformer;

- relate the number of turns in the primary and secondary of a transformer to the voltage across each, and given three of the above quantities, calculate the fourth;

- apply the principle of conservation of energy to a transformer, showing that power cannot be increased;

- identify the effect of self-induction in a coil;

- distinguish between the effect of an inductor on a high frequency AC and a low frequency AC.

A CURRENT-CARRYING WIRE IN A MAGNETIC FIELD

1.   We considered the coil in an electric motor to act like a small magnet that is pulled around by a large permanent magnet surrounding it. A simple example of the same phenomenon is shown in Figure 16.1. Here a wire passes between the poles of a horseshoe

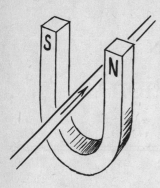

Figure 16.1

magnet. A current is in the wire in a direction away from the reader. If you try this, you will find that there is an upward force on the wire, tending to pull it out of the magnet. (The effect is fairly weak unless you have a strong magnet and a current of at least a few amps.)

(a)   In what direction would the force be if the current were reversed? _____

(b)   What would happen if you reversed the current and also reversed the magnet? _____

(a) downward (into the magnet); (b) upward (just as before the change)

2.   The directions involved in the example of frame 1 can best be remembered by use of a memory device involving the right hand. Hold the thumb and first finger of the right hand so that they are mutually perpendicular, as shown in the figure at the left. (When the thumb points up and the middle finger points to your chest, the index finger will point to your left— try it.) Now let the *M*iddle finger represent the *M*agnetic field, the *I*ndex finger represent the current (symbolized by an *I*), and the *TH*umb represent the direction of the force, or *TH*rust. Use this rule to verify the direction of the force on the wire in Figure 16.1. The wires of the coil on a motor are acted on by this same force and can be analyzed in like manner.

Now refer to Figure 16.2. The magnetic field is represented by X's (and is caused, perhaps, by a large magnet which is now shown).

Figure 16.2

(a)   In what direction is the magnetic field in relation to the page? _____

(b)   Use the right-hand rule. What is the direction of the force on the wire? _____

(a) into the page (Chapter 15, frame 4); (b) down the page

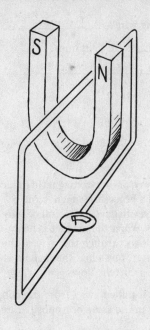

INDUCING A CURRENT

3.   The discovery (early in the 19th century) that a magnetic field could exert a force on a current-carrying wire led to the speculation, and eventual discovery, that a magnetic field could produce a current. The figure at the left shows a wire placed between two poles of a (strong) magnet, the ends of the wire being connected to a meter that can detect small currents. If you try this you will find that no current flows while the wire is stationary with respect to the magnet. If the wire is moved up or down through the magnetic field lines, however, the meter will indicate that a current is flowing.

To explain this we will return to the phenomenon of a force on a current-carrying wire. The second figure on this page shows a section of a wire in a magnetic field (enlarged from Figure 16.2), and we imagine that we are able to see a single electron in the wire.

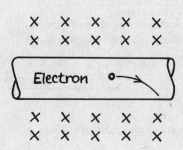

(a)   Which way is the electron moving?

_____

(b)   What is the direction of flow of conventional current?_____

(c)   What is the direction of the force on the electron due to the magnetic field?

_____

– – – – – – – – – – – – – – – – – –

(a) to the right; (b) to the left (Chapter 14, frame 2); (c) downward (Don't forget that the rule refers to *current* flow.)

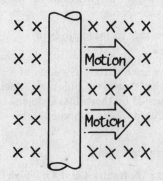

4.   The electron in our wire is unable to leave the wire. This creates a force on the wire itself. This is the reason that we observe the force on a current-carrying wire in a magnetic field.

Now consider a wire without a current being moved across a magnetic field. Each electron in the wire is being moved through the field. (See the figure at the left.) Just as in the last figure, the electron feels a downward force through the wire. Thus we see that this "induction" of a current by a wire moving relative to a magnetic field is not a completely separate phenomenon, but an extension of what we had seen previously.

(a)   Which way does conventional current flow through the wire?_____

(b)   What seems to be the effect of moving a wire across magnetic field lines?

_____

––––––––––––––––––––

(a) up; (b) create, or induce, a current

A COIL AND A MAGNET

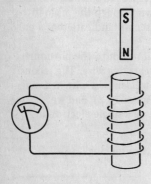

Figure 16.3

5.   The actual discovery of electromagnetic induction was not made with a horseshoe magnet and a single wire, but with a bar magnet and a coil of wire, as shown in Figure 16.3. When the magnet is held stationary, no current flows through the wire of the coil and through the meter, but when the magnet is pushed toward the coil, a current flows.

(a)   When the magnet is pulled away from the coil, does current flow in the same or opposite direction?_____

(b)   What would you expect to happen if you add more turns of wire to the coil and moved the magnet near it? _____

_____

––––––––––––––––––––

(a) the opposite direction; (b) more current will flow, because more parts of the wire cut across lines of force

LENZ'S LAW

6.   Lenz's law tells us in what direction current will flow when a magnet is moved near a coil.

   The induced current flows in such a direction so that its magnetic field opposes the change which produced the current.

When the magnet is at a position shown in Figure 16.3 there are relatively few lines of magnetic field through the coil but when the magnet is pushed closer, the number of lines through the coil increases. The direction of the field represented by these lines is downward (away from the magnet's north pole). A current is induced so as to oppose this increase in the number of lines.

(a)   So in what direction will the field caused by the current be? _____

(b)   Use the "curled fingers" right-hand rule (from Chapter 15, frames 4 and 6). Which way will current flow across the side of the coil facing you to produce this field?

_____

––––––––––––––––––––

_____

(a) upward; (b) left to right

7.   Let's work through another example. Suppose a south pole is pulled away from the top of the coil in Figure 16.3.

(a)   Does the magnetic field in the coil increase or decrease as the magnet is pulled away?

_____

(b)   What is the direction of this field?_____

(c)   In what direction will the induced current seek to establish a field? _____

(d)   How must the induced current flow to get this field? _____

– – – – – – – – – – – – – – – –

(a) decrease (It is this *decrease*, rather than the existence of the field, which causes the induced current.); (b) upward; (c) upward; (d) left to right across the front of the coil

TWO COILS: THE TRANSFORMER

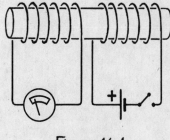

Figure 16.4

8.   If two coils are wound side-by-side and current is suddenly caused to flow in one of them, a short-lived current will be induced in the other. This results from the fact that when current starts to flow, a magnetic field is being established where none existed before. Likewise, when an existing current in one of the coils is turned off, a short-lived current is induced in the other.

In Figure 16.4, suppose the switch is closed, allowing direct current from the battery to flow in the right coil.

(a)   Which direction will current flow in the wire on your side of the coil? _____

(b)   In the center of the coil, what will be the direction of the magnetic field?

_____

(c)   What will be the direction of the field which the current induced in the left coil

seeks to produce? _____

(d)   What direction does current flow in the left coil?_____

– – – – – – – – – – – – – – – –

(a) up; (b) right to left; (c) left to right (Lenz's law tells us that the magnetic field of this induced current must oppose the buildup of the field toward the left.); (d) down (in the wires on your side)

9.   A steady direct current in one coil does not induce a current in the other because there is then no changing magnetic field.

(a)   After the switch in Figure 16.4 is closed for a while, in what direction will current

flow in the left coil? _____

(b)   When the switch is then opened, stopping the current in the right coil, what is the

direction of the current induced in the left one? _____

— — — — — — — — — — — — — — —

(a) it won't flow at all; (b) up through the wires on your side of the coil (It tends to keep
the field from decreasing.)

10.   Recall that the magnetic field of an electromagnet can be increased by placing a
piece of iron down the center of the coil. The effect in frames 8 and 9 can also be increased by placing a piece of iron through the coils. Figure 16.5 shows this iron in the
form of a ring, which causes the magnetic field to be almost entirely restricted to the centers of the coils, where it will have the most effect. Such a device is called a *transformer.*

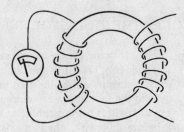

Figure 16.5

Suppose an alternating current is caused to flow in the right coil.

(a)   What is the effect of the constantly increasing and decreasing current on the mag-

netic field produced by the right coil?_____

(b)   What is the resultant effect in the left coil? _____

(c)   Suppose direct current is caused to flow in the right coil. What will be the resulting

effect in the left coil? _____

— — — — — — — — — — — — — —

(a) it constantly varies; (b) it induces a constantly varying current; (c) no current flows,
except when the DC is turned on and off, since the magnetic field doesn't change

VOLTAGE AND POWER THROUGH A TRANSFORMER

11.   Suppose more turns of wire are placed on the secondary coil (the one at left in
Figure 16.5) than on the primary (through which current first flows). Each wire of the
coil contributes to the induction effect, so the voltage across the ends of the secondary
coil will be more than across the primary. In fact, the ratio of voltages is equal to the ratio
of turns in the coil:

$$\frac{V_s}{V_p} = \frac{N_s}{N_p}$$

where V represents voltages in the secondary and primary, and N represents the number of turns in each coil. This is the main function of transformers: to increase or decrease the voltage of alternating current. The large box high on the pole on your street changes the high voltage (perhaps thousands of volts) of the transmission lines to the 110 and 220 volts used in your house.

(a)   Suppose the voltage in the lines is 1100 V. Would the number of turns in the secondary to produce 220 volts there be $\frac{1}{5}$ that of the primary, $\frac{1}{10}$, or $\frac{1}{20}$?

_____

(b)   If a transformer has 150 turns in the primary and 600 in the secondary, what voltage must be applied across the primary to produce 200 volts across the secondary?

_____

— — — — — — — — — — — — — —

(a) $\frac{1}{5}$; (b) 50

12.   Does the fact that voltage can be increased mean that you can increase power output by the use of transformers? Of course not. Remember that power is the product of voltage and current. If a transformer has twice as many turns in the secondary as in the primary, the voltage is doubled, but the amount of current available to be drawn from the secondary is decreased to one half. Thus the product of voltage and current remains the same. We can write the relationship between the currents I and voltages V in the primary and secondary as:

$$I_p V_p = I_s V_s$$

Since the product of current and voltage is equal to power (Chapter 14, frame 21), this equation tells us that the conservation of energy is not violated; there is no more energy flowing from the secondary than flows into the primary.

(a)   Two amps at 120 volts flow through the primary of a transformer. If the secondary voltage is 12,000 volts, what current can be drawn from it?_____

(b)   One amp at 240 volts flows through the primary of another transformer. If the primary has 500 turns and the secondary 50, what current can be drawn from the secondary? _____

— — — — — — — — — — — — — —

(a) 0.02 amps; (b) 10 amps ($V_s = 24$; $I_p V_p = I_s V_s$)

SELF-INDUCTION

13.   We know that the buildup of magnetic field in a coil causes a current to be induced in the coil. But what about the primary coil, the one which is causing the magnetic field in the first place? The field is building up in that coil also. Theoretically, the magnetic field buildup should cause an induced current in that coil, just as it did in the secondary coil. This tendency, called self-induction, does exist.

Refer back to Figure 16.4. Consider what happens when the switch is closed. Current starts to flow. As soon as it does, the growing magnetic field tries to start a current in the opposite direction. There is, in effect, a "back voltage" placed on the coil which opposes the voltage producing the current. Which of the following would seem to result?

_____ (a)   No current flows.

_____ (b)   Current increases rapidly to compensate.

_____ (c)   Current does not increase as rapidly as expected.

– – – – – – – – – – – – – – – –

c is what actually happens

14.   Recall that it is the changing of the current, rather than the existence of the current, which causes induction. Because alternating current is by its nature constantly changing, a coil of wire impedes its flow. Alternating current which is changing direction with greater frequency causes more induction in a coil, and thus a coil impedes the flow of high frequency AC more than it does low frequency AC. Devices called inductors, consisting of wire wrapped around metal cores, are used to allow low frequencies to pass but impede high frequencies.

(a)   Will an inductor offer any impedance to the flow of DC? _____

(b)   Which of the following frequencies will flow best through an inductor: 60 cycles/

second, 300 cycles/second, or 1000 cycles/second? _____

– – – – – – – – – – – – – – – –

(a) no; (b) 60 cycles/second

SELF-TEST

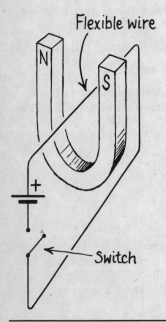

Flexible wire

N

S

+

Switch

The questions below will test your understanding of this chapter. Use a separate sheet of paper for your diagrams or calculations. Compare your answers with the answers provided following the test.

1.   What will happen when the switch is closed in

the circuit of the figure at the left? _____

_____

_____

2.   Suppose a current runs down a wire from the top to the bottom of this page, and that a magnetic field runs through the page from the other side toward you. What will be the direction of

the force on the wire? _____

3. Suppose a wire carries a current from east to west along the equator, where the magnetic field is from south to north (geographic directions—recall that the magnetic poles are near the "wrong" geographic poles). What is the direction of the force on the wire? _____

4. Refer to the first figure in frame 3. What will happen if the magnet is quickly moved upward (while the wire remains stationary)? _____
    _____

5. Suppose the north pole of a magnet is pulled away from the lower end of a coil which is wound like that of Figure 16.3. In which direction will the induced current flow through the wires which can be seen on the side of the coil facing you?
    _____

6. Why is an iron core included in transformers? _____
    _____

7. Why can a transformer not be used to transmit direct current? _____
    _____

8. Can a transformer be used to increase the power output of an electric circuit? Why or why not? _____
    _____

9. A transformer has 500 turns in the primary and 3000 in the secondary. If 120 volts AC is sent across the primary, what voltage will appear across the secondary?
    _____

10. If 2.0 amps flow through the primary of the transformer of question 9, what is the maximum current which can be drawn from the secondary?_____

Answers to Self-Test

If your answers do not agree with those given below, review the frames indicated in parentheses before you go on to the next chapter.

1. The wire will move (downward) between the poles of the magnet due to the force on the current-carrying wire.   (frames 1, 2)

2. toward the left (Use the right-hand rule.)   (frame 2)

3. downward, toward the ground (right-hand rule)   (frame 2)

4. A current will be induced in the wire.   (frame 3)

5.   from left to right (As the magnet is pulled away, the field—which is upward—
      decreases in the coil. The induced current opposes this decrease. Thus it is such as
      to cause an upward magnetic field. The curled fingers right-hand rule of the last
      chapter tells us that it must move across the visible wires from left to right.
      (frame 6)

6.   The iron core increases the magnetic effect by adding its own magnetic field.
      (frame 10)

7.   A change in magnetic field through the secondary is necessary to induce a current
      in it. A direct current will produce a constant field rather than a changing one.
      (frame 9)

8.   No. Although voltage can be made to increase, current will be decreased in the same
      proportion.   (frame 12)

9.   720 volts   (frame 11)

10.   0.33 amps   (frame 12)

CHAPTER SEVENTEEN

Electromagnetic Waves

Prerequisites: Chapters 13 (frames 6–8), 14–16

The link between electricity and magnetism on the one hand and light on the other is not an obvious one. Sure, we all flick an electric switch to turn on a light, but it is electromagnetic fields which provide the real link between the two branches of physics. We will see that light can be considered as waves from electric and magnetic fields. The details of this connection will be saved until the next chapter; right now we will spend most of our time on another type of electromagnetic wave—the radio wave.

When you have completed your study of this chapter, you will be able to:

- describe how a changing electric field is generated by an antenna;

- specify the condition of an antenna when its electric field is at a maximum and when its magnetic field is at a maximum;

- identify a factor important in the construction of a transmitting radio antenna which is related to the frequency of the wave;

- differentiate between the processes of AM and FM radio transmission;

- identify, in order of frequency, a number of portions of the electromagnetic spectrum;

- describe two methods of detecting an electromagnetic wave;

- relate the number on a radio dial to the radio wave which is received;

- relate light to electromagnetic waves.

PRODUCTION OF RADIO WAVES

1. Consider two wires connected to a source of alternating voltage as in figure A on page 188. The source causes electrons to move up into the upper wire, then down into the lower, then back up, and so on. Figure A shows the device at the instant when the top of it is positive and the bottom is negative. At this instant, there is an electric field (represented by the arrows) in the region immediately surrounding the device.

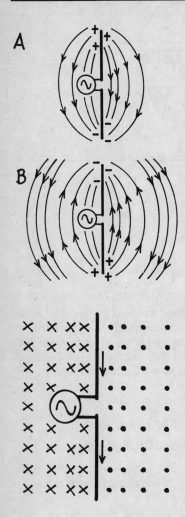

A

B

A moment later, the ends of the wire switch electric polarity, and the situation of figure B occurs.

(a) In what direction does the electric field near the wire point now? _____

(b) Has the previous downwardly-directed field disappeared? (Refer to the figure.) _____

(c) Where has it gone? _____

_____

(a) upward; (b) no; (c) it has moved farther from the device

2. The field continues to spread out from the source so that an electron located some distance from it would feel a force first upward, then downward, then upward, and so on. (Recall that the electron, having a negative charge, feels a force in a direction opposite to the direction of the field.) An electric wave has been sent through the space between the device and the electron feeling the force. In fact, if this electron is one of the many located in a receiving antenna (which is very similar to the device illustrated in frame 1 except that a current detector replaces the AC source) the electric wave can be detected at great distances.

But the story does not stop here. The figure at the left shows the device—actually a transmitting antenna—during the time when the electrons are in motion from the bottom to the top of the wires. Since a current is flowing in the wires, a magnetic field surrounds them. The field is shown into the paper on the left side and out of the paper on the right because the conventional current is downward and the right-hand rule gives this as the correct direction of the field.

Current flows down the wire until the bottom is positive and top negative (at which time the electric field is at a maximum), but then it starts flowing upward, producing a magnetic field in a direction opposite to that of the figure in this frame. Just as the electric field spread out from its source after being produced, a magnetic field detected at some distance from the transmitting antenna will be first in one direction and then in the other. It will alternate direction with the same frequency as the electric field and the same frequency as the alternating current in the antenna. Since a changing magnetic field causes a current to flow in a coil of wire, such a coil is effective in detecting the magnetic wave.

Neither an electric wave nor a magnetic wave can exist without the other; their combination is called an electromagnetic wave.

(a) What two components make up an electromagnetic wave? _____

_____

(b)   Describe what is happening in a transmitting antenna when it produces a maximum electric field. _____

(c)   When does it produce the maximum magnetic field?_____

_____

(d)   What is necessary to induce a current in a coil of wire?_____

_____

(a) electric wave and magnetic wave; (b) one end is positive and the other is negative; (c) when the current in it is at a maximum; (d) a changing magnetic field

3.   One might think that the frequency with which electromagnetic waves are sent out depends only upon the frequency of alternation of the AC source. In practice, however, the length of the antenna is important in getting maximum wave amplitude. Just as a strip of metal (like a saw blade) which is clamped at one end has a natural frequency at which it will vibrate back and forth, there is a natural frequency with which electrons alternate back and forth in an antenna. The value of this frequency is determined by the length of the antenna. To make an antenna which will transmit a wave of maximum amplitude, it is constructed such that its frequency of natural electron vibration corresponds to the AC frequency.* Antennas such as the one shown in frame 2 send out radio waves, which have frequencies from a few cycles per second to about a billion cycles per second. This is a far greater range than the radio waves we normally think of. AM radio uses a range from about 530 to about 1600 kilocycles per second, while FM radio frequencies range from about 88 to 108 million cycles per second. We use the remainder of the radio band for such purposes as citizen's band radio, television, short wave radio, and microwave ovens.

(a)   What affects the amplitude of the transmitted wave besides the voltage of the AC

source? _____

(b)   Name four uses of various parts of the radio section of the electromagnetic spectrum.

_____

Figure 17.1

(a) antenna length; (b) AM radio, FM radio, television, citizen's band radio, shortwave radio, microwave ovens

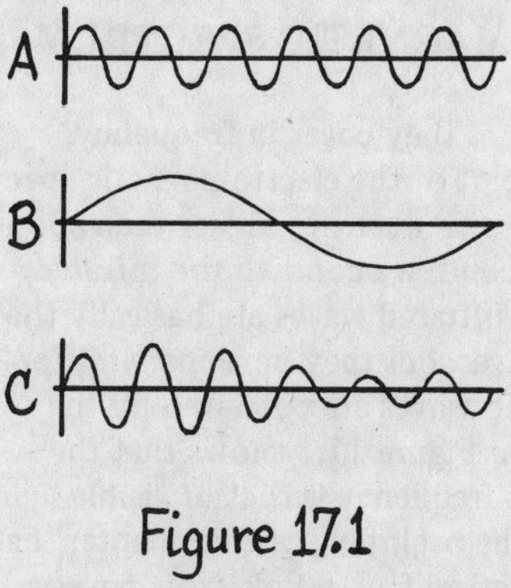

4.   If the electromagnetic wave described in frame 1 were represented graphically, it would look like wave A in Figure 17.1. The graph simply tells us the wave exists and that it is a smooth, regular wave. In AM radio, a signal is put on the wave by *Amplitude Modulation*. For example, suppose wave B represents a sound wave which the local disc jockey wishes to

_____

*If you studied Chapter 11 you will recognize such frequency-matching as the condition necessary for "resonance."

transmit. This wave is used to "modulate" the basic radio wave (called the carrier wave) so that the transmitted wave looks like wave C. The amplitude of the sound wave has been used to control—to "modulate"—the amplitude of the carrier wave. Look carefully to see the similarity of pattern between graphs B and C of the figure.

In tuning your radio receiver, you adjust the frequency of your radio to the frequency of the carrier wave (so that only the desired station is received), and your radio set separates the sound wave from the carrier wave and reproduces the former as sound.

What does "amplitude modulation" do to the radio wave? _____

_____

_ _ _ _ _ _ _ _ _ _ _ _ _ _ _

It changes the amplitude of the carrier wave according to the sound frequency.

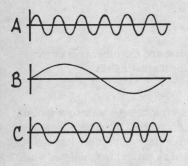

Figure 17.2

5.   FM stands for *Frequency Modulation*. In FM radio, it is not the amplitude of the wave which is modulated, but the frequency. Wave A in Figure 17.2 is a carrier wave (unmodulated). Wave B of the figure is a sound wave which is used to modulate the carrier wave. The modulated wave, C, has been changed in frequency so that it carries the sound signal. Although the electronic details of the transmitter and receiver differ, the principle is the same as for AM radio.

What is the basic difference between AM and FM radio signals? _____

_____

_ _ _ _ _ _ _ _ _ _ _ _ _ _ _ _

In AM the amplitude of the carrier wave is modified; in FM the frequency is modified.

THE ELECTROMAGNETIC SPECTRUM

6.   As great a range as they cover in frequency, radio waves are only part of the electromagnetic spectrum. As we advance to frequencies higher than about a billion cycles per second, we come to the infrared part of the spectrum. Infrared waves are basically the same type as radio waves, but they have many different properties. Infrared waves are perceived by the body as heat radiation. Figure 17.3 shows that the region of next highest frequency is that of visible light. It was mentioned at the beginning of this chapter that light and electromagnetism have a link. You can see from Figure 17.3 that light acts as an electromagnetic wave.

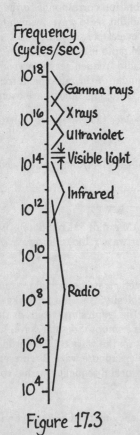

Frequency (cycles/sec)

Figure 17.3

We said earlier that the size of the transmitting antenna is related to the frequency of the wave produced. Infrared waves cannot be produced by wire antennas because these antennas cannot be made short enough for the high frequency of infrared. An extremely short antenna would be needed—in fact, about the size of an atom. It is vibration of the atom which produces infrared radiation. And to produce light, the vibration must occur within the atom itself. The production of light and its properties are the subjects which we will take up in the next few chapters.

(a)   Refer to Figure 17.3. What types of waves have greater frequencies than light?

_____

(b)   What is the highest frequency for which antennas can be made (approximately)?

_____

_ _ _ _ _ _ _ _ _ _ _ _ _ _ _ _

(a) Ultraviolet, x-rays, gamma rays; (b) 10^{12} to 10^{13} cycles/second

SELF-TEST

The questions below will test your understanding of this chapter. Use a separate sheet of paper for your diagrams or calculations. Compare your answer with the answers provided following the test.

1.   On the illustration of a transmitting radio antenna at the right, show the arrangement of electric charges when a maximum downward electric field is being transmitted.

2.   What is the condition of the electrons in the transmitting antenna when

maximum magnetic field is being transmitted? _____

_____

3.   How can the electric field portion of an electromagnetic wave be

detected? _____

_____

4.   How can the magnetic field portion of an electromagnetic wave be detected?

_____

5.   Besides the frequency of the alternating current source, what else is important in

determining the frequency of a transmitted electromagnetic wave? _____

_____

6.   Choose some number on an AM radio dial and explain what is meant by that num-

ber. _____

_____

7. What is meant by amplitude modulation? _____

_____

8. Explain what frequency modulation means. _____

_____

9. List in order of frequency, from low to high, at least four portions of the electro-

magnetic spectrum. _____

10. What basic connection is there between light and electromagnetism?_____

Answers to Self-Test

If your answers do not agree with those given below, review the frames indicated in parentheses before you go on to the next chapter.

1. Positive charges are at the top of the antenna and negative charges at the bottom.
    (frame 1)

2. A maximum current is flowing, so the electrons have maximum speed up or down
    the antenna. The electric current produces the magnetic field.   (frame 2)

3. It can be detected by an antenna similar to the transmitting antenna except that a
    detector of electric current replaces the voltage source.   (frame 2)

4. The magnetic wave is best detected by placing a loop of wire (with its ends hooked
    to a current detector) in the path of the wave. The changing magnetic field causes a
    current in the loop.   (frame 2)

5. the length of the transmitting antenna   (frame 3)

6. For example, "900" on the dial means that a station transmitting with a carrier fre-
    quency of 900,000 cycles/second is received at that point on the dial. (Some dials
    may show this as "90" or "9.")   (frame 3)

7. In amplitude modulation, the amplitude, or intensity, of the carrier wave is changed
    according to the frequency of the sound being transmitted.   (frame 4)

8. In frequency modulation, it is the frequency of the carrier wave which is changed,
    or modulated, slightly according to the frequency of the sound being transmitted.
    (frame 5)

9. The ones listed in this chapter, from low to high frequency: radio, infrared, visible,
    ultraviolet, x-rays, and gamma rays.   (frame 6 and Figure 17.3)

10. Light acts as an electromagnetic wave.   (frame 6)

PART FIVE

The Physics of Light and Optics

Part V of this book is the longest, for it considers the many faces of the phenomenon of light. It has been mentioned earlier in the book that light acts as an electromagnetic wave, but we will find that things are not quite as simple as we would like, for under certain circumstances light acts as though it is definitely *not* a wave. Three chapters will be needed to uncover the various aspects of this paradox.

In addition, we will study some of the more practical manifestations of light, such as reflection and refraction (the bending of light as it travels from one material into another). This will allow us to look at optical instruments varying in complexity from magnifying glasses and eyeglasses to microscopes and telescopes. Part V ends with a study of the nature of color and the manner in which colors combine to form other colors.

CHAPTER EIGHTEEN
Light: Wave or Particle?

No Prerequisites, but Chapters 10–12 recommended

The nature of light has been a puzzle down through the ages. Early theories included the idea that the eye sends out particles which go to the thing seen. In newspaper cartoons even today you sometimes see little arrows pointing from a character's eye to an object to illustrate that the character is looking at the object. Eventually people asked the question, if this is the nature of light, why can't we see in the dark? As we will see, however, the real conflict historically has been between theories which demand much more serious treatment than this primitive one.

A study of this chapter will enable you to:

- specify why Galileo failed to measure the speed of light with his experiment;
- describe how Roemer used the period of one of Jupiter's moons to measure the speed of light;
- specify how Michelson measured the speed of light on earth;
- state the speed of light in metric and British units;
- specify why Newton clung to a particle theory of light;
- identify the scientist who discovered that light definitely acts as an electromagnetic wave and state his method of discovery;
- interpret the chart of the electromagnetic spectrum;
- interpret the observations of photoelectric effects based on wave theory;
- explain why Michelson and Morley expected to be able to detect the ether;
- state evidence for and against both the wave and the particle nature of light.

GALILEO AND THE SPEED OF LIGHT

1.   One of the problems in analyzing light is simply to measure its speed. In about the year 1600, Galileo Galilei tried to do this. He stationed himself on a hill and used a lantern to send a flash of light to a friend on another hill a few miles away. When the friend saw

the flash from Galileo's lantern, he would answer with a flash from his. All Galileo would have to do was to measure the time from when he exposed his lantern to when he saw the return flash and then subtract his friend's reaction time. As you might suspect, all Galileo could conclude was that light moved at a speed faster than he could measure. He could not even rule out the possibility that it moves instantly from place to place—with infinite velocity.

Which of the following could explain Galileo's results?

_____ (a)   Measurement of time had not progressed enough to measure the speed of light.

_____ (b)   His lanterns were too close together.

_____ (c)   His lanterns were too far apart.

————————————————

Choice a (In fact, it was Galileo who invented the pendulum clock. Answer b may be considered a possible answer, but there could be no hills far enough apart to allow the measurement without the ability to measure extremely short times.)

ROEMER'S MEASUREMENT

2.   The first successful measurement of the speed of light was reported by Olaus Roemer in 1675. His method was based on the astronomical observation that Io, the innermost moon of the planet Jupiter, has a period of revolution around the planet of 42.5 hours.* Once Roemer observed it crossing in front of Jupiter, he could predict that it would again eclipse Jupiter 42.5 hours later. And every 42.5 hours after that. In fact, he could predict accurately at what time of night Io would eclipse Jupiter three months or six months later.

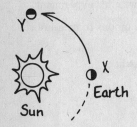

Figure 18.1

Consider Figure 18.1. When the earth is at point X, Roemer observes Io eclipsing Jupiter. He does some calculations and predicts that three months later this should happen again at a certain time of night. However, when he makes his observations three months later, he finds the measurement is wrong. What is the most likely explanation of this?

_____ (a)   The period of Io changed.

———————————

*You may be interested to know that it was Galileo who discovered that Jupiter has moons—the first, other than our own, ever seen.

___(b)   The earth is farther from Jupiter.

___(c)   Jupiter is in a different position.

_ _ _ _ _ _ _ _ _ _ _ _ _ _

b (The period of a moon doesn't change measureable in three months, and Jupiter does not move far enough in that time to make a difference.)

3.   When the earth was at position Y (in Figure 18.1) Roemer found that the eclipse happens 500 seconds later than predicted at X. This unexpected result is not due to an irregularity in the orbit of Io, but is due to the fact that light from Jupiter requires 500 seconds longer to reach earth at position Y than it did when the earth was at X.
    Roemer did not know the earth-to-sun distance as accurately as we do now, so the value he obtained was not as accurate as we can calculate. His observation and calculation, however, formed the first successful measurement of the speed of light. Let's use our known distance from the sun ($1.5 \cdot 10^{11}$ meters)* and divide by the time (500 seconds) to find

the speed of light. _____

_ _ _ _ _ _ _ _ _ _ _ _ _ _

$3 \cdot 18^8$ m/s (or 186,000 miles/second)

MICHELSON'S METHOD

4.   Late in the nineteenth century, Albert Michelson devised an earthbound method to measure the speed of light. Figure 18.2 shows light hitting an eight-sided mirror, bouncing over to a regular mirror (M), bouncing back to hit another side of the eight-sided mirror, and finally bouncing into Michelson's eye at the bottom. Now suppose that the octagonal mirror is rotated clockwise. Flashes of light will now be sent down to mirror M. When the rotating mirror is at the position shown in the figure, light will leave it in the right direction to hit M and be returned again. But by that time, side Z (where it should hit) will

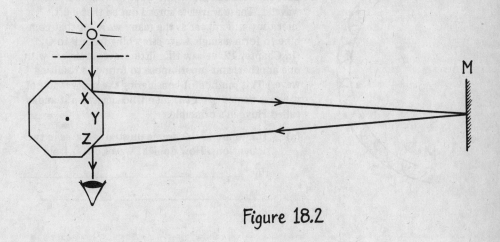

Figure 18.2

---

*This is equivalent to 93,000,000 miles, the average radius of the earth's orbit.

have moved around a little, so that the light flash will not be bounced into the eye. So Michelson speeds up the rotation. He speeds it up until, while the light flash is going from X to M and back, side Y has moved over to replace Z in its position. The light then reflects from Y to the waiting eye. All that Michelson needed to know was the rotational speed of the mirror (from which he could calculate time) and the distance to mirror M.* Distance divided by time yielded for Michelson the speed of light.

(a)   What was the advantage of Michelson's calculation over that of Roemer? _____

_____

(b)   If Michelson had used a twelve-sided mirror rather than an eight-sided one, would

he have to rotate it faster or could he have rotated it more slowly? _____

— — — — — — — — — — — — — — —

(a)   Michelson could measure his variables more accurately, and he wasn't dependent on astronomical measurements.
(b)   more slowly (Only $\frac{1}{12}$ of a revolution would have been required while the light was traveling.)

HUYGENS' PRINCIPLE

5.   Early experimenters showed that light does not travel at an infinite speed, and that its speed can be measured. But what is light? The ancient view was that it is a stream of particles, or corpuscles, which leave the object being viewed and travel to the eye. In the mid-1600s Robert Hooke, and English physicist, proposed that light may be a wave. This theory was then improved by Christian Huygens, who explained that light waves leaving a source could be considered to be the result of a number of tiny wavelets, each emitted

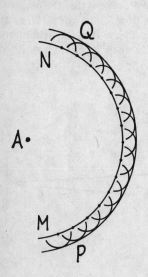

by each point on the wave. Thus light leaving point A in the figure at the left consists of little wavelets which at some time later appear as the wave from M to N. To see how the wave spreads from there, suppose that each point on the wave MN emits a little wavelet. These wavelets spread out so that a little later they will appear as the many wavelets that combine to form a single wave stretching from P to Q. (In Chapter 12 we saw that little waves *interfere* with one another, and superimpose to form the resulting wave.) This method of considering the progression of waves is named after the man who suggested it and is called Huygen's principle.

(a)   The figure in this frame illustrates waves in two dimensions. How do light waves differ from

this? _____

_____

_____

*The distance to mirror M was more than 20 miles in the original experiment.

(b)   In two dimensions, each wavelet forms part of a circle. What is formed in three

dimensions? _____

— — — — — — — — — — — — — — — —

(a) they spread in all directions—in three dimensions; (b) part of a sphere

NEWTON'S PARTICLES

6.   As you probably know, Isaac Newton is one of the giants—perhaps *the* giant—of
classical physics. Newton was a contemporary of Robert Hooke, but he could not accept
Hooke's wave theory of light. Newton realized that if light is a wave, it should bend
around corners. If any such bending occurs, it is not obvious.

Newton had done experiments with light and had observed that it broke into colors
when passing through a prism, but that the separated colors themselves could no longer
be broken further into more colors. His theory was that each color corresponds to a certain type of particle, or corpuscle. White light, according to this theory, is composed of
particles of all types. Each type passes through the glass of a prism at a different speed,
and this causes each to bend a different amount. Thus the light separates into different
colors. He also observed that when all of the colors are bent back together again, white
light results. This fit the theory well.

Since Newton had such a great reputation, the fact that he favored a particle theory
of light may have resulted in slowing the progress of work toward finding the true nature
of light. After all, if the master says that light is a particle, why experiment and theorize
in any other direction? Which of the following did Newton use as evidence that light acts
as a particle rather than as a wave?

_____ (a)   Light doesn't seem to bend around corners.

_____ (b)   The speed of light can be explained by considering it a particle.

_____ (c)   Light is emitted by hot objects.

_____ (d)   The separation of light into colors can be explained with a particle model.

— — — — — — — — — — — — — — — —

a and d

ELECTROMAGNETIC WAVES

7.   In the early 1800s Thomas Young performed an experiment which showed conclu-
sively that light has wave characteristics. (If you have studied Chapter 12, you may want
to jump ahead to frames 1 through 4 of Chapter 22 to see the nature of this experiment.)
But it was not until the middle of the 1800s that the nature of the wave was discovered.
James Clerk Maxwell was doing theoretical work on electricity and magnetism at the time.
This work led him to predict the possibility of the existence of electromagnetic waves—
waves given off when electrical charges are vibrated in a wire. Maxwell formulated a set of
equations which, when the waves were observed 30 years later, were found to fit their
behavior quite well. In the course of his work, he did a calculation of the predicted speed
of the electromagnetic waves. The speed his calculation showed was $3.0 \cdot 10^8$ meters per
second. He realized, of course, that this was the same as the speed of light, and since this

was not likely to be just a coincidence, he concluded that light is a form of electromagnetic wave. We know now that all electromagnetic waves move at this speed.

(a)   Who first showed experimentally that light acts as a wave? _____

(b)   Who first concluded that light was an electromagnetic wave? _____

— — — — — — — — — — — — — — —

(a) Thomas Young; (b) James Clerk Maxwell

8.   You may know that the frequency of sound waves determines the pitch of the sound we hear. In the case of light, the frequency of the light wave determines the color of the light seen. The waves in the visible spectrum have frequencies between about $4 \cdot 10^{14}$ cycles/second and $7.5 \cdot 10^{14}$ cycles/second. (This is from 400,000,000,000,000 to 750,000,000,000,000.)* Red light has the lowest frequency. Then comes orange, yellow, green, blue, and finally violet at the highest frequency.

      The electromagnetic spectrum ranges far beyond the range of visible light, with frequencies far higher and far lower. Figure 18.3 illustrates this range. Just beyond the violet, we find ultraviolet and the highest frequency waves are gamma rays. At the low end, the spectrum ranges from radio waves to infrared.

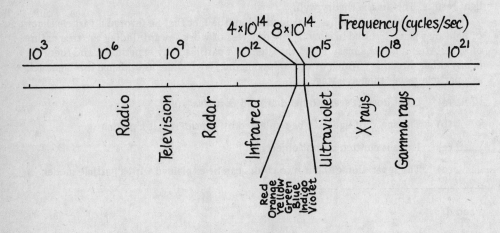

Figure 18.3

(a)   About what frequency of electromagnetic wave is used for television? _____

(b)   What is the speed with which x-rays travel? _____

— — — — — — — — — — — — — — —

(a)   about 10^9 cycles/second
(b) $3 \cdot 10^8$ m/s (Or 186,000 miles/s. All electromagnetic waves travel at the same speed; see frame 7.)

_____

*See Appendix I for a discussion of powers of ten notation.

THE PHOTOELECTRIC EFFECT

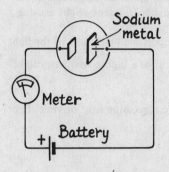

9.   Maxwell's work with the theory of electromagnetic waves may seem to have solved the problem of the nature of light, but at least one major problem remained. There was one experiment—the photoelectric effect—which could not be explained by considering light as a wave. Suppose we set up an electric circuit as shown in the figure at the left. In this circuit the negative terminal of a battery has been connected to a piece of sodium metal. The positive terminal of the battery is connected through a meter which measures electric current to another piece of metal a centimeter or so from the sodium. Both of these metal plates, as they are called, are enclosed in an airtight glass tube from which the air has been removed. With the experimental set-up as described, no current flows from the battery and the electric meter shows a reading of zero. The thing not mentioned so far in this set-up is light. If light is shined onto the sodium plate, electric current will flow and register on the meter. If you block the light hitting the sodium plate, the electric current stops. And, as one might suspect, when the amount of light hitting this negative plate is increased, the amount of current is increased.

But this is still not the whole story. It turns out that if one tries various colors of light, he finds that violet and blue light are able to cause current to flow, but colors toward the red end of the spectrum do not result in electric current when shined on the sodium. Somehow, the frequency of the light is important, with higher frequency light causing current and lower frequency light unable to do so.

In Chapter 14 we saw that electric current in a wire consists of electrons flowing from the negative end of the wire to the positive end. In the circuit of interest here, no electrons were able to flow when no light hit the sodium because the electrons were unable to jump across the vacuum between the negative and positive plates. Somehow the light is able to bump electrons out of the negative plate. Then once they are free from that plate, they are attracted to the positive plate and they continue on through the meter to the positive terminal of the battery.

(a)   What is needed to cause current to flow in the circuit shown in this frame?

_____

(b)   What component of the circuit emits electrons when light hits it? _____

_____

— — — — — — — — — — — — — — — —

(a) high frequency light; (b) sodium metal

10.   The observations of the photoelectric effect are summarized below:

(1)   Current flows as soon as light hits the negative plate. No "build-up time" is needed, no matter how dim the light.

(2)   High frequency light causes electrons to be emitted from the sodium, but low frequency light does not. (The critical frequency below which electrons will not be emitted depends upon the metal used as the negative plate. The vast majority of metals require light of frequencies well above the visible region of the spectrum.)

(3)   Once the frequency of light is great enough to cause electron emission, the energy of the emitted electrons does not depend upon the brightness of the light. It depends upon the light's frequency, higher frequency light causing higher energy electrons to be emitted.

(4)   The amount of current which flows depends upon the brightness of the light.

(a)   Which will cause more current to flow, a brighter light or a higher frequency light?

_____

(b)   Which would produce emission of higher energy electrons, blue light or violet light?

_____

– – – – – – – – – – – – – – – –

(a) brighter; (b) violet

ELECTROMAGNETIC THEORY
AND THE PHOTOELECTRIC EFFECT

11.   Because of the success of Maxwell's electromagnetic theory, at the end of the 1800s light was considered to be wavelike in nature. Let us see how well this theory explains the photoelectric effect. Read the statements below, relate each to observations in the last frame, and indicate which statements seem to support the wave theory.

_____ (a)   The energy contained in electromagnetic waves and the amount of that energy which would strike an individual electron can be calculated. Such a calculation shows that an electron could indeed gain enough energy to break free from sodium, but only after light was shined on it for some hours (the amount of time depending upon the brightness of the light).

_____ (b)   Since light waves carry energy whether they are of high or low frequency, the frequency of light should have no effect on whether or not electrons are emitted from the metal.

_____ (c)   The frequency of the light wave should not affect the emission of electrons. If anything, one might expect the amplitude of the wave to affect the energy of the emitted electrons.

_____ (d)   Brighter light carries more energy and might be expected to eject more electrons.

– – – – – – – – – – – – – – – –

Statement d (Since one in four is not nearly enough, the photoelectric effect was a major roadblock in the way of total acceptance of the wave theory of light.)

THE ETHER

12.   Another problem with the electromagnetic theory arose over the question of the ether. This ether (sometimes spelled "aether") has nothing to do with the ether gas we use today in medicine; it is the medium through which electromagnetic waves were expected to pass. Every other wave has some medium through which it passes. Water waves require a water surface; waves going down a stretched spring obviously require a spring;

and sound waves likewise require a material (usually air) for their propagation. The question arose as to what medium propagates light waves. This medium must possess some unusual characteristics.

> It must exist throughout space, all the wave to the stars. The earth, in its journey around the sun, is continually moving through the ether, but it apparently passes through it without friction. No material known to man would allow such motion through it, but the ether must somehow be able to.
> Since light passes through some solids, such as glass, the ether must actually permeate these transparent substances.

The ether, then, must not be a *material* substance at all. It must not be made up of atoms and molecules.

(a)   Why must the ether extend to the stars? _____

(b)   Why must the ether not be a material substance? _____

_____

– – – – – – – – – – – – – – –

(a)   light reaches us from the stars
(b)   It must permeate material such as glass and water. In addition, the earth must pass through it without friction.

THE MICHELSON-MORLEY EXPERIMENT

13.   The speed of sound *relative to the air* is 1100 feet/second. Consider a man at the center of a railway flatcar, as shown in Figure 18.4 If the car is standing still and there is no wind, a shout from the man will be heard at the same time by a person at each end of the car.

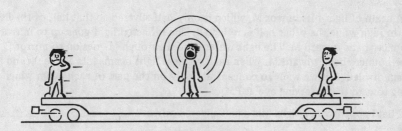

Figure 18.4

But what happens when the car is in motion, say, 60 ft/s (about 40 mi/hr)? In this case air is moving by the shouting man at 60 ft/s. When a person shouts, he does not *throw* the sound as one throws a baseball; he simply disturbs the air and the disturbance moves away through the air at 1100 ft/s. But if the air is moving toward the shouter at 60 ft/s, then the sound actually moves away from the man toward the front of the railway car at 1040 ft/s. And relative to the car, it moves toward the person at the rear at the rate of 1160 ft/s. If the person at the rear did not take into account that the car was

moving, what would he measure the speed of sound to be?_____

– – – – – – – – – – – – – – – –

1160 ft/s

14.   Now let's consider light and the ether. The velocity of light is $3 \cdot 10^8$ m/s (or
186,000 mi/s). The earth, however, must be moving through the ether in much the same
way as the railway car is moving through the air. If this is the case, the speed of light
should be slightly more or less, depending upon which direction the light is traveling
through the ether. The earth's speed around the sun is about $3 \cdot 10^4$ m/s (20 mi/s), very
slow compared to the speed of light. Thus the difference in the speed of light due to the
motion of the earth should be very difficult to detect.
        In 1881, Albert Michelson and Edward Morley decided to try to detect the ether.
To perform this very difficult measurement, Michelson and Morley used the experimental
set-up illustrated in the figure below.

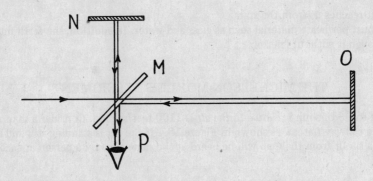

        A beam of light hits mirror M, which is only half-silvered so that half of the light
passes through while the other half is reflected. The reflected light goes up to mirror N
and bounces back toward M. The light which passed through M goes on to mirror O and
likewise bounces back toward M. When each of these light beams hits M the second time,
it is again divided, but we want to concentrate only on the part of each beam which ends
up going toward the observing eye at P.

(a)   What happens when light hits a half-silvered mirror?_____

_____

(b)   What path did the light follow in getting from the source to the eye in the Michelson-

Morley experiment?_____

_____

– – – – – – – – – – – – – – – –

(a)   half goes on through and half is reflected
(b)   to mirror M, then half to N and half to O, then both halves back to M, then to the
        eye.

15.   Suppose each of the beams of light travels exactly the same distance from the time it leaves M till it returns to M. In that case, a crest of the wave which went one direction will exactly meet the crest of the wave going the other way. Likewise, troughs of the two waves will meet at M. In Chapter 12 we saw that this caused constructive interference. If you did not study that chapter, suffice it to say that when crest meets crest and trough meets trough, the light which went one way will add to and reinforce the light which went the other way.

On the other hand, if the light travels just slightly slower along one path, a crest from one direction will land on a trough from the other, resulting in destructive interference. This means that the two beams of light will cancel one another out.

In actual practice, Michelson and Morley did not have to set the paths to exactly the same length. They set the apparatus up so that they saw a pattern of interference when they made an observation with the apparatus at a fixed position. Then they rotated the entire set-up by 90 degrees and they looked for a change in the pattern. Why did they expect to see a change? Suppose that as they originally started the experiment, the ether was flowing past the earth in the direction from mirror N to mirror M. The light going from M to N was thus fighting against the "ether wind." In going back to M, it was moving with the ether wind. Light making the trip from M to O and back, on the other hand, was continually moving in a direction *across* the wind. Analysis of such motions predicts that the speed across the wind will be greater than the average speed upwind and downwind. Now the experimenters rotate the appartus. Light which was once going across the wind now travels up and down, and light which traveled up and down is now going across. Thus, no matter how the two distances compare, there should be a *change* in the interference pattern when the apparatus is rotated. To make a long story less long, no change was seen. The experiment was tried again and again, and calculations (which indicated that such a change should show up) were checked and rechecked. The result was the same: no change in the interference pattern. Two conclusions were possible: (1) maybe the ether travels along with the earth so that it causes no ether wind, or (2) there is no ether. The first of these is ridiculous—we found out long ago that the earth is not at the center of the universe. We can't expect ether to follow the earth around. We have to conclude that there is no ether. And we are back to the problem of what light travels in if it is a wave. The only answer that we can give, even today, is that electromagnetic radiation is a vibration of the electromagnetic field, which is not a material substance. Light is the "force-signal" sent by an electric charge through this field.

(a)   Why was it not necessary for each light path to be exactly the same length in the

experiment?_____

_____

(b)   What was the overall purpose of the Michelson-Morley experiment?_____

_____

— — — — — — — — — — — — — — —

(a)   Michelson and Morley were looking for a *change* as they rotated the apparatus.
(b)   to detect the ether (or the earth's motion through the ether)

SUMMARY

Is light a wave or a particle? We have examined some of the evidence pertaining to this question and have concluded that it doesn't seem to be either. Maxwell's electromagnetic theory seems to explain most of the observations very well, but it falls short on the experiment concerning the photoelectric effect. In addition, we have to conclude that there is no substance which carries the wave.

But why do we expect light to be *either* a wave or a particle? We expect this because we cannot imagine anything else it could be. But that does not mean that it cannot be something else. There is no question but that light has wave aspects—it exhibits the phenomenon of interference, and that cannot be explained except with a wave explanation. But it also exhibits particle effects in the photoelectric effect—it acts as if particles are knocking electrons out of a metal. Perhaps we cannot say that it *is* a wave or that it *is* a particle, but must live with the fact that whatever it *is*, it sometimes acts like a wave and sometimes acts like a particle. We will see in the next chapter that there is a theory which, at least in part, describes light by uniting these two aspects of nature.

SELF-TEST

The questions below will test your understanding of this chapter. Use a separate sheet of paper for your diagrams or calculations. Compare your answers with the answers provided following the test.

1. What did Galileo conclude concerning the speed of light? _____
    _____

2. What causes the apparent variation in the period of the innermost moon of Jupiter?
    _____
    _____

3. In Michelson's method for measuring the speed of light, how did he determine how long the light took to go to the distant mirror and return to his eight-sided mirror?
    _____

4. What is the speed of light? _____

5. Did Isaac Newton favor the wave or particle theory of light? _____

6. What did Newton propose as the cause of light having various colors? _____
    _____

7. What was surprising about the speed Maxwell predicted for electromagnetic waves?
    _____

8. What spectral color has the highest frequency? What type waves have a frequency just higher than this? (Refer to Figure 18.3.) _____

9.   Light necessary to cause emission of electrons from sodium metal has a fixed (minimum/maximum)_____ frequency.

10.   What does wave theory predict concerning the time required for electrons to be emitted once light is shined on a metal? _____

_____

11.   What effect does wave theory predict that the frequency of light will have on the emission of electrons from metals? _____

_____

12.   The Michelson-Morley experiment was an attempt to determine the earth's speed through what? _____

13.   What was concluded from the Michelson-Morley experiment?_____

_____

Answers to Self-Test

If your answers do not agree with those given below, review the frames indicated in parentheses before you go on to the next chapter.

1.   It was faster than he could measure, and perhaps infinitely fast.   (frame 1)

2.   The apparent variation results from the fact that at some times the earth is farther from Jupiter than at other times, and it takes light longer to get to us when we are farther away.   (frames 2–3)

3.   This time was equal to the time for his mirror to rotate through $\frac{1}{8}$ turn. He had to know the frequency of rotation of the mirror.   (frame 4)

4. $3 \cdot 10^8$ m/s or 186,000 mi/s   (frame 3)

5.   the particle theory   (frame 6)

6.   Newton proposed that each color corresponds to a certain type particle. (frame 6)

7.   It was the same as the known speed of light.   (frame 7)

8.   violet; ultraviolet   (frame 8 and Figure 18.3)

9.   minimum   (frame 11)

10.   Some time should be required while the electron builds up energy.   (frame 11)

11.   It should not affect the emission of electrons.   (frame 11)

12.   ether   (frame 14)

13.   There is no evidence for the existence of an ether.   (frame 14)

CHAPTER NINETEEN
The Quantum Nature of Light

Prerequisites: Chapter 5 and Chapter 18, frames 8–10

In the last chapter we saw that many aspects of the behavior of light can be understood when you consider it to be a wave. Some of its behavior, however, demands a particle explanation. In this chapter we will discuss more of these phenomena and will present a view of light which—at least partly—unites the wave and particle nature of light.
   When you complete your study of this chapter you will be able to:

- relate blackbody radiation to the temperature of the object;

- identify the problem concerning the relationship between Maxwell's electro-magnetic theory and blackbody radiation;

- specify Planck's contribution to the study of the nature of light;

- relate Planck's formula to the wave and quantum theories of light;

- specify Einstein's contribution to the theory of light;

- interpret the photoelectric effect with regard to the quantum theory;

- relate the Bohr model of the atom to the quantum theory of light;

- determine the amount of energy emitted when an electron de-excites, or the energy needed to excite;

- differentiate between excited and stable states in an atom;

- relate the frequencies of emitted photons to the use of spectra in identifying elements;

- name at least two ways of exciting electrons;

- differentiate between emission spectra, absorption spectra, and the continuous spectrum;

- explain the difference in functioning of an incandescent and a fluorescent lamp;

- compare what happens when electrons in a solid are excited to those in a gas;

- distinguish between fluorescent and phosphorescent materials;

- explain how "black light" posters work;

- state the difference between laser light and regular light;

- specify the derivation of the word "laser."

BLACKBODY RADIATION
AND ELECTROMAGNETIC THEORY

1.   If you turn on a burner of an electric stove to its lowest setting, you can feel heat radiated from the burner even though it does not glow visibly. As you adjust the burner to a hotter setting, you can feel more heat radiated and you begin to see the burner glowing red. With a yet higher setting, the burner goes from red to bright yellowish-red. Figure 19.1 is a graph of the intensity of radiation emitted by an object at various temperatures. The line corresponding to 6000°C shows most of the radiation being emitted at a higher frequency than do the lines for 4000°C and 3000°C.

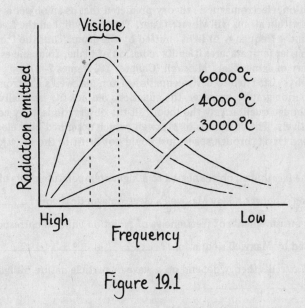

Figure 19.1

The heat you feel radiated near the stove burner is infrared radiation, which is of lower frequency than visible light. When the burner was set at low, it was not hot enough to emit visible light. As the temperature was increased, however, it started emitting light of higher and higher frequency until red light was emitted. As it continued getting hotter, it emitted more light toward the higher frequency end of the spectrum, causing the appearance of an orangeish- or yellowish-red. If you are able to continue increasing its temperature, the burner would become white in appearance. This is exactly what happens to the filament of a light bulb.

Such radiation from a solid is referred to (in the ideal case) as blackbody radiation because, it turns out, a perfectly black object is the best radiator of electromagnetic energy.

(a)   Would an object at a temperature of 3500°C emit more or less radiation than it would at 3000°C? _____

(b)   How does the frequency of light emitted depend upon the temperature of the object? _____

(c)   Would a black or chrome electric burner radiate heat more efficiently?_____

— — — — — — — — — — — — — — — —

(a) more; (b) as the temperature rises, the maximum frequency becomes greater; (c) black

2.   How well did electromagnetic theory explain the radiation curves from black bodies? This theory predicted that when an object gets hotter, it emits more radiation. And this is indeed the case, as shown in Figure 19.1, for the curves for higher temperatures rise higher and have more area under them. (The area under the curve shows the total amount of radiation of all frequencies emitted.)

        In addition, electromagnetic theory predicted that as an object is heated, the electrical charges within atoms will vibrate faster. The theory linked the frequency of this vibration with the frequency of light emitted by the atom. Thus the fact that as an object is heated to higher temperatures it emits radiation of higher frequencies also seems to fit with the thoery of James Clerk Maxwell (Chapter 18, frames 7–8).

        When physicists tried to derive equations from Maxwell's electromagnetic theory to fit the radiation graphs, however, the equations did not fit, especially at lower frequencies. Thus although qualitatively the theory fit the observations, they could not be verified quantitatively. The equations of Maxwell, which described how electromagnetic waves are emitted and travel through space, just could not be fit to the situation of blackbody radiation.

(a)   Does the fact that an object heated to a higher temperature emits higher frequencies of radiation support or deny Maxwell's electromagnetic theory? _____

(b)   Do the measurements of frequencies of radiation emitted correspond with those predicted by Maxwell's equations?_____

(c)   Does Maxwell's theory depend on a wave or particle nature of light? _____

— — — — — — — — — — — — — — — —

(a) support (at least qualitatively); (b) no; (c) wave

PLANCK'S QUANTUM HYPOTHESIS

3.   In 1900, Max Planck, a physicist at the University of Berlin, presented an explanation for the behavior of black bodies. We saw earlier that as the frequency of vibration of the electric charges within atoms increased, the frequency of radiation emitted was expected to increase. It was expected that any frequency of vibration was possible, just as you can swing a stick above your head at any frequency you want (up to the limit of your physical abilities, anyway). Planck, however, proposed that the electrons in atoms are limited to certain specific energies of vibration. It is as if you could swing the stick at frequencies of 3 swings per second and 4.5 swings per second, but were unable to swing it

at any frequency in between. Planck stated that the atoms could exist only in these specific energy states, and when they absorb and release energy they change from one energy state to another. The energy absorbed is thus absorbed in little chunks, or *quanta.* Likewise, he said that when energy is emitted, it is emitted only in quanta. In addition, the energy contained in each quantum is governed by the frequency of the radiation. The formula for the relationship is:

$$E = h \cdot f$$

where E is the energy of a quantum of radiation of frequency f which is absorbed as the black body radiates away energy. The symbol h is called Planck's constant and is a number we use as a multiplier of frequency to calculate the amount of energy in a quantum.

(a)   According to Planck's theory, can an atom gain or lose energy in any amount?

_____

(b)   How does the quantum theory differ from the pure wave theory of light? _____

_____

(c)   In Planck's formula, what is f the frequency of? _____

_____

— — — — — — — — — — — — — — —

(a) no, only in discrete chunks; (b) no continuous flow; (c) the photon of radiation absorbed or emitted by an object

4.   The radiation curves shown in Figure 19.1 can be derived mathematically using Planck's quantum hypothesis. Planck's proposal, however, was radical for physics of his day. Early in the history of our understanding of nature we wondered whether or not *matter* was of such a nature that it could be divided into smaller and smaller bits forever without coming to some smallest chunk. We know today that it cannot. There is a smallest chunk into which any substance, at least in principle, can be divided. It is the atom or molecule. Energy, however, had always been considered to be continuous, to occur in any amount, with no limit to how small an amount of energy could be exchanged between two objects. Planck's theory put energy into the same category as matter, for he stated that energy was also "chunky."

(a)   How did Planck see energy differently from other sicentists of his day? _____

_____

(b)   How does Planck's theory of energy relate to our current theory of matter?

_____

— — — — — — — — — — — — — — —

(a)   He saw it as existing in quanta, rather than continuous.
(b)   Both can be divided into smallest "pieces."

EINSTEIN ON THE PHOTOELECTRIC EFFECT

5.   Physicists recognized in 1900 that Planck's hypothesis did indeed explain blackbody radiation, but it was not quickly accepted as being the true nature of things. The quantum idea did not receive much attention until 1905 when Albert Einstein accepted the hypothesis and took it one step further. Planck held that radiation was absorbed and emitted in quanta, but he clung to the idea that radiation itself was continuous. Thus he pictured Maxwell's electromagnetic waves traveling through space as before, but being emitted and absorbed in small quanta.

   Einstein proposed that radiation actually travels through space as quanta of energy. This was another blow to the electromagnetic theory. One now had to somehow think of little chunks of energy traveling along instead of smooth, steady waves. Einstein called the chunks "photons," and we will use "photon" interchangeably with "quantum" in the rest of this book. The wave theory was not abandoned. In fact, the amount of energy in the photons was determined by the frequency of the radiation! As we saw above in Planck's equation, the energy was still directly related to the radiation frequency.

(a)   Write Planck's equation. _____

(b)   How did Einstein's idea of how energy traveled differ from Planck's? _____

   _____

   _____

   — — — — — — — — — — — — — —

(a) $E = h \cdot f$; (b) Einstein thought energy traveled in photons, while Planck thought it traveled in smooth waves.

6.   Four observations about the photoelectric effect were presented in frame 10 of the last chapter. We will review them here from the standpoint of the quantum theory (but we will take them in a different order). In 1905 Einstein used Planck's ideas to explain the photoelectric effect. According to Einstein, an electron is ejected from an atom (in the photoelectric effect) when it absorbs a photon of sufficient energy to allow it to free itself from the atom. Instead of gradually gaining energy from electromagnetic waves, the energy transfer is done all at once in quantum jumps. The four observations:

(1)   Electrons are emitted immediately when light strikes the photoelectric surface. According to quantum theory, since electrons gain their energy in single steps rather than gradually gaining energy from a wave, this observation holds true. As soon as a photon of sufficient energy strikes an electron, the electron jumps from the metal.

(2)   High frequency light causes electrons to be emitted, while low frequency light does not. According to quantum theory, photons of high frequency light have more energy than photons of low frequency light. Since some minimum energy is needed to free electrons from atoms, apparently photons of low frequency light do not have sufficient energy.

(3)   The energy of emitted electrons depends upon the frequency of the light striking the metal. Again, this is easily explained. Light of high frequency has photons of great energy, so when one of these photons is absorbed by an electron, the energy left over above what is needed to free the electron goes toward giving the electron extra speed.

(4)   According to quantum theory, brightness of light depends upon the number of photons in the beam rather than the amplitude of a wave. Thus when the light is brighter, more photons cause more electrons to be emitted.

Thus the photoelectric effect, which was one of the major obstacles to full acceptance of the electromagnetic theory of light, was fully explained by the quantum theory. How does the quantum theory explain each of the following:

(a)   Ultraviolet light causes electrons to be emitted while infrared light does not.

_____

_____

(b)   A bright blue light causes more electrons to be emitted than a dim one.

_____

_____

— — — — — — — — — — — — — — —

(a)   Photons of ultraviolet light have more energy since ultraviolet light is of a higher frequency than infrared.
(b)   Bright light has more photons than dim light, and each photon causes an electron to be emitted.

THE BOHR MODEL OF THE ATOM

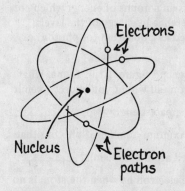

Electrons

Nucleus

Electron paths

7.   The figure at the left shows the common representation of the atom as many people think of it today. In the Bohr model, the atom consists of negatively charged electrons revolving around the central nucleus. Most of the mass of the atom by far is contained in the nucleus, the electrons accounting for less than $\frac{1}{4000}$ of the mass of the atom.

Bohr realized, as did the science community around 1900, that the electromagnetic theory, when applied to the nuclear atom, predicted that electrons should continually be emitting radiation because their orbiting motion is essentially a vibration, and a vibrating electric charge should emit electromagnetic radiation. (This idea was at the heart of the electromagnetic theory and had been verified time and time again. Radio transmission depends upon this idea.) If the atom emitted radiation as predicted, it would gradually lose energy so that the electron would gradually fall in toward the nucleus. The atom would die! Obviously this does not happen.

To solve the problem, Bohr made two assumptions about the atom, and in the process, he explained how photons of light come into being. The two assumptions: (1) electrons in orbit around the nucleus orbit in specific given orbits, and (2) no radiation is emitted when they are in these orbits. Since each orbit corresponds to a certain energy, the energy levels of an atom are not continuous. When an electron jumps from one allowed orbit to another, the atom absorbs or emits radiation.

(a)   For an electron to move to an orbit of greater energy, must it lose or gain energy?

_____

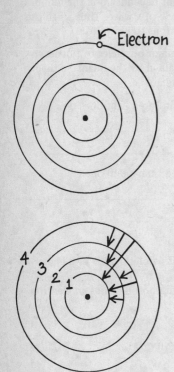

(b)   When it falls to an orbit of lower energy, does it gain energy or emit it as radiation? _____

– – – – – – – – – – – – – – – –

(a) gain; (b) emit

EMISSION OF LIGHT

8.   The figure at the top left represents an atom with four energy levels for the electrons. (In reality an atom has many more allowed orbits than this.) The figure shows one electron in the outer orbit. No electrons are in the inner orbits. An atom with an electron in anything but its lowest possible orbit has more than a minimum amount of energy and is said to be *excited*. This is not a stable condition for an atom to be in, and the atom will release its extra energy by means of the electron jumping—or rather falling—down to a lower level.

The second figure in this frame shows the possible falls that the electron might make. It can fall to levels 3, 2, or 1. If it first falls to 3, it can then fall to 2 or 1. Once in level 2, it will then fall on down to 1. Each time the electron falls from an outer orbit to an inner one, it emits an amount of energy which corresponds to the difference in energy of the levels. This energy is released as a quantum (or photon) of radiation.

(a)   Examine the second figure in this frame. What possible electron fall would you expect to emit the photon of greatest energy? _____

(b)   What is the maximum number of photons that this one excited atom could emit?_____

(c)   Where will the electron be when the atom is no longer excited?_____

– – – – – – – – – – – – – – – –

(a) 4 to 1; (b) 3 (4 − 3, 3 − 2, 2 − 1);
(c) in the inner orbit

9.   Figure 19.2 is a representation of the energy levels of a hydrogen atom. Notice that levels of energy rather than circular orbits are shown. The units of energy filled in here are electron volts (eV); don't worry about what the unit means just now. In addition, the energy is negative to conform to standard

Figure 19.2

practice in physics. Suppose the electron is in the −0.85 eV level and falls to the −1.51 eV level. The electron loses energy since the level to which it is falling is lower than its original level. The amount of energy lost is the difference between energy levels: −0.85 − (−1.51) = 0.66 eV. Thus the photon emitted has an energy of 0.66 eV.

(a)   Suppose the electron falls from the −0.85 eV level to the −3.40 eV level. How much energy will the emitted photon have? _____

(b)   If it falls from the −0.85 eV level to the lowest level, how much energy will the photon have?_____

(c)   Suppose that after the electron falls from −0.85 to the −1.51 eV level, it then falls to −13.6 eV. How much energy will the emitted photon have?_____

— — — — — — — — — — — — — — —

(a) 2.55 eV; (b) 12.75 eV; (c) 12.09 eV

10.   The energy levels in Figure 19.2 are, in fact, the energy levels of hydrogen atoms. Thus when hydrogen atoms emit radiation, the radiation is always of certain characteristic frequencies. Other elements have different sets of energy levels. No two elements have the same set of energy levels. Thus no two elements emit photons of the same set of energies.

Now recall the tie-in of the quantum theory with the electromagnetic theory. If the energy of two photons is different, the two photons differ in frequency. If we limit ourselves to considering visible light, we see that light of different frequencies appears as different colors. Thus we find that each element in nature, if heated up to the gaseous state so that it emits light, will emit its own characteristic frequencies, its own characteristic colors.

(a)   Could elements be uniquely identified by the color of light their gases emit?_____

(b)   Why or why not? _____

_____

— — — — — — — — — — — — — — —

(a) yes; (b) each element has a different set of energy levels, so a different set of frequencies would be emitted

11.   Now suppose we take the light emitted by heated hydrogen gas and pass it through a prism so that it breaks down into its colors. Only certain frequencies of light are present in light emitted by hydrogen atoms, and therefore only certain colors result. The spectrum of light produced from hydrogen atoms looks somewhat like that of Figure 19.3. Note that some of the "colors" of light are in the ultraviolet region. Although these frequencies are invisible to the eye, they can be recorded on film.

If a different chemical element is used as the source of light, a different spectrum is produced, because each element has its own set of energy levels for its electrons, and thus its own characteristic frequencies of emitted light. The characteristic spectrum of helium and sodium gas are also shown in Figure 19.3. Thus if upon heating a gas you see the particular combination of frequencies which correspond to those of helium, as in the figure, you may be sure that the light was emitted by the element helium.

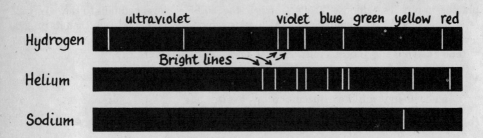

Figure 19.3

The type spectrum discussed above is, for obvious reasons, called an *emission spectrum* or a *bright line spectrum*. When light coming from a glowing gas is separated into its colors (by using a prism, for example), one can determine what elements are present in the gas. This analysis of the spectrum of various materials is done with an instrument called a spectroscope, and the technique is called spectroscopic analysis.

(a) Refer to Figure 19.3. What element shown has the most ultraviolet light in its

spectrum? _____

(b) What color light does sodium emit? _____

(c) Why might you want to do a spectroscopic analysis? _____

_____

– – – – – – – – – – – – – – – –

(a) hydrogen; (b) yellow; (c) determine what element or elements make up a substance

EXCITATION OF ATOMS

12. We saw how excited atoms lose their extra energy. But how do they get excited in the first place? The simplest way is by heating. When a gas is heated, the atoms move more and more rapidly and collisions become more violent. When collisions become violent enough that there is sufficient energy available to cause an electron to jump to a higher level, it does.

The heating may occur because of a chemical reaction occurring among the atoms. This is what is happening in a flame. The heat being produced by the burning is causing the atoms of the gas to become excited. After excitation, the electrons quickly fall back to an inner orbit and give off their extra energy as light (in the form of photons).

Atoms may also be excited by an electric current passing through a gas. Electric current in a gas is simply electrons and positively charged atoms rushing through the gas. When these fast moving particles collide with the atoms of the gas, they give them energy, exciting them.

Name two ways of causing atoms to absorb energy and become excited. _____

_____

– – – – – – – – – – – – – – – –

heating, electric current

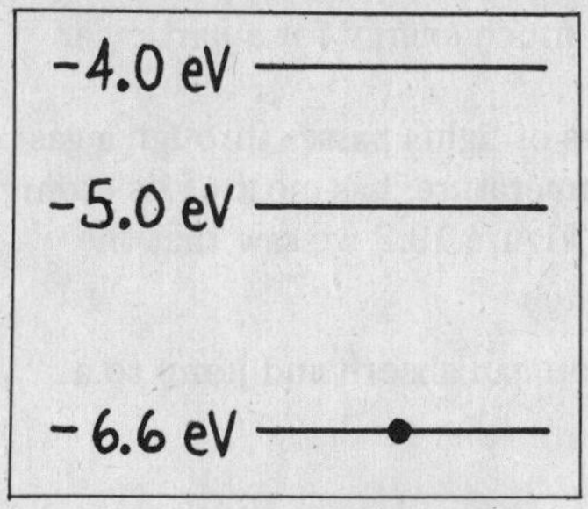

```
-4.0 eV  ————————
-5.0 eV  ————————

-6.6 eV  ———•———
```

Figure 19.4

13.   Let us look at what happens when atoms are excited by fast moving electrons. Figure 19.4 is the energy level diagram for an imaginary atom with an electron in a low level and with two allowed levels of energy above it.

(a)   In order to jump to the next highest level, how much energy must the electron gain? _____

(b)   To jump from that position to the second excited state, how much energy must it gain?

_____

(c)   How much energy will be released when it finally returns to the stable state?

_____

— — — — — — — — — — — — —

(a) 1.6 eV; (b) 1.0 eV; (c) 2.6 eV

14.   Suppose an electron (part of an electric current) is moving with an energy of 1.4 eV and strikes the orbiting electron of Figure 19.4 in its stable state. Since the approaching electron does not have enough energy to knock the orbiting electron all the way up to its first excited level, it cannot move it at all, and it will therefore not affect it. But suppose it has 1.8 eV of energy. In this case, it will give up 1.6 eV of energy upon striking the orbiting electron, causing this electron to jump up to the next level. The striking electron is left with 0.2 eV of energy, and moves on at a reduced speed (corresponding to the 0.2 eV of energy). If the incoming electron has an energy of 2.6 eV or higher, it can cause the orbiting electron to jump up to its second excited level.

   Once the orbiting electron has been given energy and is in a high energy state, it can then fall back down to its original state. In so doing it will release a quantum of energy, as we have seen.

   Suppose our stable atom is struck by one electron after another. (Though this is unrealistic, it will emphasize a point.) The approaching electrons have the energies below. What will happen in each case? (Consider this to be a sequence of events.)

(a)   1.5 eV _____

(b)   2.0 eV _____

(c)   0.5 eV _____

(d)   1.0 eV _____

— — — — — — — — — — — — —

(a) no effect; (b) electron jumps to −5.0 eV level, and striking electron continues with 0.4 eV of energy (moving slower); (c) no effect; (d) electron jumps to next level, and striking electron stops

15.   Another way to excite the atoms of a gas is by allowing them to absorb photons of light. This will happen if the photon has just the right amount of energy to cause the electron to make the jump to a higher level. If the photon has too little energy, the

electron cannot make the jump. But if the photon has too much energy for a particular jump, the electron will not make that jump either!

Suppose white light (which consists of all frequencies of light) passes through a gas of atomic hydrogen. Any gas, unless it is at a very high temperature, has most of its atoms in the lowest possible energy state, the "ground" state. In Figure 19.2 we saw that the hydrogen electron has a ground state energy of -13.6 eV.

(a)   What is the least amount of energy which this electron can absorb and jump to a

higher orbit?_____

(b)   What are three other photon energies that could cause hydrogen atoms to become

excited?_____

— — — — — — — — — — — — — — — — —

(a) 10.2 eV (13.6 − 3.4); (b) 12.09, 12.75, and 13.15 eV (actually all four could happen simultaneously in separate atoms)

16.   Now remember that we said that we were shining white light through the gas. The gas, however, is absorbing out those frequencies that correspond to jumps which its electons can make. Thus the light which passes through the gas will lack the frequencies that correspond to the photons absorbed. The spectrum of light getting through is similar to the figure in this frame. This figure shows a continuous spectrum containing all frequencies except those absorbed. Such a spectrum is called an *absorption*, or *dark line*, *spectrum*. (Recall that a heated gas produces a bright line spectrum.)

Since the frequencies of absorption are fixed by the possible electron jumps, the dark lines appear at particular characteristic positions in the spectrum—the same positions where the bright lines appeared in the emission spectrum of that particular chemical element. Thus an absorption spectrum can also be used to identify elements. The light coming from stars passes through the nonglowing gases near the star as well as through the earth's atmosphere before reaching us. Therefore it appears as an absorption spectrum. If we eliminate from consideration those absorbed frequencies which are due to the earth's atmosphere, we can examine the spectrum of light from a star and determine what chemical elements are in the star's atmosphere. This is a standard spectroscopic procedure in astronomy.

(a)   How does an emission spectrum compare to an absorption spectrum for the same

element? _____

_____

(b)   Why do the dark lines of the absorption spectrum appear at the same place where

the emission spectrum has bright lines? _____

_____

— — — — — — — — — — — — — —

_____

(a)   The emission is bright for certain frequencies while the absorption is dark for the same frequencies.

(b)   They result from electrons jumping between the same levels. (In one case the electron jumps "up" and in the other case it falls "down.")

TYPES OF LAMPS

17.   A regular incandescent light bulb, which is made with a filament of tungsten, emits white light, rather than frequencies characteristic of the element tungsten. Since the tungsten filament is a solid, rather than a gas, the result is different. In a solid, the individual atoms are not independent of one another, but instead are linked closely together. Thus the electrons of an atom are not influenced exclusively by their "home" atom. In fact, electrons jump from one atom to antoher in metals and have no "home." As a result, a photon of essentially any energy can be emitted in one electron jump or another. The outcome of this is a *continuous* spectrum, containing all the colors of the rainbow.

(a)   In a solid are there more or fewer possible energy jumps? _____

(b)   What color light emission results when electrons are not restricted as to energy levels? _____

(c)   If the solid is not hot enough to produce the high energy photons at the blue-violet end of the spectrum, what color would you expect it to emit? _____

_ _ _ _ _ _ _ _ _ _ _ _ _ _ _ _

(a) more; (b) white; (c) red (see frame 1)

18.   Some street lights have a more bluish tint than others. These lamps are most likely mercury vapor lamps. As their name suggests, they contain a vapor of the element mercury. When the mercury is heated by an electric current, it emits light. The mercury vapor lamp does not show a continuous spectrum, but instead shows an emission (bright line) spectrum. And the frequencies in this spectrum are characteristic of the element mercury.

In some parts of the country you see bright yellow street lamps. These are probably sodium vapor lamps, containing the vapor of the element sodium. This element, when heated to a vapor form, emits only yellow light. (See Figure 19.3 for the sodium spectrum.)

(a)   Why does a gas vapor lamp not emit white light as does the incandescent lamp?

_____

_____

(b)   Our eyes are more sensitive to yellow light than any other color. What advantage does this suggest for sodium vapor lamps? _____

_____

_ _ _ _ _ _ _ _ _ _ _ _ _ _

(a)   Gas vapor lamps produce emission spectra of certain frequencies, while solids in incandescent lamps don't.

(b)   Since sodium vapor produces only yellow light, they appear brighter to us than other lamps of the same wattage. (In fact, they are 6 to 8 times more efficient than incandescent lamps.)

19.   If you pass the light from a fluorescent tube through a prism and observe its spectrum, you will see a combination of a bright line spectrum and a continuous spectrum. The bright line spectrum is easy to explain—the tube contains gas, primarily mercury. To see why the continuous spectrum also shows up, we must consider the phenomenon of fluorescense.

Most solids, when light shines on them, will either reflect the light or absorb it and then emit the same frequency they absorbed. Certain substances, called fluorescent, have the property of being able to absorb high frequency light—high energy photons—and then emit light of a lower frequency. Such a fluorescent substance is used to line the inside of the tube of the fluorescent lamp. Some of the frequencies emitted by mercury gas are in the ultraviolet region of the spectrum. Suppose the atom of the figure in this frame absorbs a photon of ultraviolet light, causing the electron to jump up to the third excited level. When the electron falls back down to a lower state, however, it can fall to any of the three states below it. And since the energy it loses if it falls to the first or second excited states is less than if it falls all the way to the ground state, the frequency associated with this fall will be less than the frequency of the ultraviolet light which started the whole process. The result is that the ultraviolet radiation emitted by the mercury causes the tube's coating to fluoresce in the visible region of the spectrum.

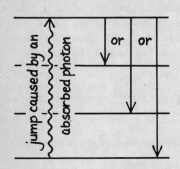

(a)   Would you expect the solid coating to emit a continuous or bright line spectrum?

_____

(b)   What causes the bright line spectrum in a fluorescent lamp? _____

(c)   Fluorescent lamps do not get as hot as incandescent lamps. Would you expect a 40 watt fluorescent lamp to emit more or less light than a 40 watt incandescent lamp?

(The wattage of a bulb tells us the electrical energy used each second.) _____

– – – – – – – – – – – – – – – –

(a) continuous (and it does); (b) the mercury vapor; (c) more (because less electrical energy is transferred to heat energy)

20.   The phenomenon of fluorescence is also responsible for the behavior of "black light" posters. "Black light" is a common term for ultraviolet light. When such a poster is exposed to ultraviolet radiation, it absorbs the ultraviolet and emits visible light. Since our eyes are not sensitive to ultraviolet, "black lights" themselves do not appear bright to us. But some of the energy they emit is absorbed and then emitted as visible light by the fluorescent dyes on the poster, so the poster itself glows.

(a)   A "black light" appears violet in color. What do you think causes this? _____

_____

(b)   How is a "black light" poster similar to a fluorescent lamp? _____

_____

– – – – – – – – – – – – – – ∟ – –

(a) visible violet light from the high end of the visible spectrum—not ultraviolet waves;
(b) both absorb high energy photons and emit lower energy ones

PHOSPHORESCENCE

21.   Some materials have the property of absorbing a photon of light but not re-emitting
the energy immediately, as if the electron jumped to an excited state and got stuck there
for a while. Suppose you have a number of atoms of such a material, and you expose
them to light. Many of them will absorb photons and become excited, and gradually,
atom by atom, they will emit visible light. You then turn off the source of light. The material will continue to glow as some of the atoms are still in the excited state and one by one
emit their extra energy in the form of visible light. Such substances are called phosphorescent, and are used for such purposes as coatings for the hands and numbers on clocks.
       Which of the common items below are phosphorescent?

_____ (a)   lighted dials on electric alarm clocks

_____ (b)   glow-in-the dark Christmas tree ornaments

_____ (c)   black light posters

_____ (d)   bright paint on a new car

_____ (e)   paint on the hands of a wristwatch

– – – – – – – – – – – – – – – – – –

b and e

THE LASER

22.   Normally, when an electron is excited to an outer orbit, it falls back to an inner
orbit, causing a photon of light to be emitted. Although the time delay between excitation and de-excitation is extremely short (except for phosphorescent substances), it does
exist. What determines how long the time will be for any individual atom? Apparently
the fall back to an inner orbit is a chance occurrence and happens spontaneously. But
there is a method by which the excited atom can be caused to de-excite and emit its photon. But before considering this, let's review briefly.

(a)   What is a material with a long delay between excitation and de-excitation called?

_____

(b)   What is a material that emits lower frequency photons on de-excitation called?

_____

– – – – – – – – – – – – – – – – –

(a) phosphorescent; (b) fluorescent

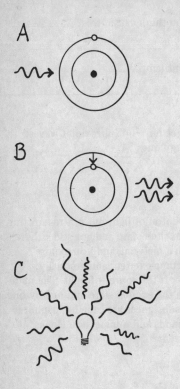

23.   Figure A at left shows an excited atom about to be struck by a photon exactly the same frequency as the photon that will be released when the electron falls inward. Such a photon will "stimulate" the atom to de-excite and emit energy. Amazingly, the emitted photon will come off in the same direction as the one causing it and in phase with that one. Figure B shows what this means. The two photons are traveling away "in step." This is the secret of the laser, the word being an acronym for *Light Amplification by Stimulated Emission of Radiation.*

Figure C shows how we might think of the waves of light from a regular incandescent lamp. Since each photon of light is emitted spontaneously, they are not in step with one another, but are all jumbled. In addition, they are of various wavelengths. They are said to be "incoherent." Laser light, on the other hand, is in phase (in step), and is of one single wavelength. It is "coherent."

(a)   Which would be easier to form into a beam, laser or incandescent light? _____

(b)   Which involves spontaneous emission of energy?

_____

(c)   What is the source of the word "laser"? _____

_____

— — — — — — — — — — — — — — — —

(a) laser; (b) incandescent; (c) *light amplification by stimulated emission of radiation*

24.   Although there are differences from one type laser to another, all have one basic feature in common: the need to have many excited atoms available to be stimulated to emit light. The laser itself may be either solid or gas. It consists of a tube containing atoms that have one excited state in which electrons tend to stick—they do not readily fall to an inner orbit. On both ends of the tube are mirrors.

To operate the laser, the atoms are excited by either a flash of light or an electric current. Electrons become fixed in the orbit referred to above until finally (actually a very short time by human standards) one of them de-excites and emits a photon. If this photon is going in any other direction except down the length of the tube, it will not cause laser action, but eventually a photon will be emitted down the tube. When it passes another excited atom, it will stimulate that atom to emit a photon and the two photons will move off in phase. Then each of them will encounter another atom. Result: 4 photons in phase, then 8, 16, 32, and so on. When they reach the end of the tube, they reflect back again and the laser action continues to grow. This would be of no value if it were confined to the laser tube, but one of the mirrors is made so that a very small fraction of the light hitting it escapes through it. This is the laser beam which today is used for everything from surveying to eye surgery.

The advantage of laser light, then, is that it is intense because of its waves being in phase. (If you studied Chapter 12 you realize that out-of-phase waves cancel one another.) In addition, the light remains concentrated—it does not spread out much as it travels along.

(a)   Will just any photon emitted in the laser tube start the laser action? _____

(b)   Where is the light emitted from the laser tube?_____

(c)   What is special about photons of laser light? _____

— — — — — — — — — — — — — — — —

(a) no; (b) through one of the mirrors on the end; (c) they are in phase and directed

SELF-TEST

The questions below will test your understanding of this chapter. Use a separate sheet of paper for your diagrams or calculations. Compare your answers with the answers provided following the test.

1.   How does the frequency of the radiation emitted depend upon the temperature of the source?_____

_____

2.   What did electromagnetic theory predict concerning the relationship between the temperature of an object and the frequency of radiation emitted? How well did the theory fit the experimental data? _____

_____

3.   Who first suggested the quantum idea which explained blackbody radiation?

_____

4.   Explain Planck's quantum hypothesis. _____

_____

5.   How did Einstein expand Planck's original quantum idea? _____

_____

6.   How does quantum theory explain the observation that high frequency light is able to cause electrons to jump free of some metals while low frequency light cannot do so? _____

_____

7.   According to quantum theory, upon what does the brightness of light depend?

_____

_____

8.   Explain, according to Bohr's model of the atom, how light is emitted by the atom.

_____

_____

9.   What type spectrum is produced by a heated gas? _____

10.   What practical application is made of the fact that the spectrum of one element is different from the spectrum of another? _____

_____

11.   Refer to Figure 19.4. Suppose an electron with an energy of 2.8 eV strikes the orbiting electron. How much energy might this electron lose in the collision? _____

12.   What is it about a fluorescent lamp which causes it to show a continuous spectrum (along with the bright line spectrum)?_____

_____

13.   What is "black light," and how does it make fluorescent paint glow? _____

_____

_____

14.   How does phosphorescence differ from fluorescence in the manner in which the atom emits light?_____

_____

15.   What causes laser light to be more intense than regular light? _____

_____

Answers to Self-Test

If your answers do not agree with those given below, review the frames indicated in parentheses before you go on to the next chapter.

1.   As the temperature of the source becomes greater, the frequency of radiation is increased.   (frame 1)

2.   Electromagnetic theory predicted that frequency would increase as temperature increases, but the prediction could not be made to fit the data in a quantitative way. (frame 2)

3.   Max Planck   (frame 3)

4.   Planck hypothesized that atoms lose and gain energy only in definite, fixed jumps, which he called quanta.   (frame 3)

5.   Einstein proposed that light itself traveled as photons (quanta) of energy. (frame 5)

6.   Low frequency light is made up of photons of low energy. Their energy may not be enough to release the electrons, but light of higher frequency has photons of higher energy and these may be energetic enough to knock electrons from atoms. (frame 6)

7.   The brightness of light depends upon the number of photons in the beam. (frame 6)

8.   Light is emitted as photons when an electron falls from an orbit of greater energy to an orbit of lesser energy. The energy of the photon is equal to the difference in energy between the two orbits.   (frames 7–8)

9.   An emission spectrum is produced by a heated gas.   (frame 11)

10.   This fact can be used to identify the elements which compose a substance. (frame 11)

11.   either 1.6 eV or 2.6 eV   (frames 13–14)

12.   The coating inside the tube is a solid, and it emits photons of many energies when it fluoresces.   (frame 19)

13.   "Black light" is the name sometimes given to ultraviolet radiation. The atoms of the fluorescent paint absorb photons of ultraviolet radiation and re-emit photons of lower energy—visible light.   (frame 20)

14.   In the case of fluorescence, the atom re-emits its energy (almost) immediately. In phosphorescence, the atom may wait quite a while before its electron returns to a lower energy level and emits the energy as light.   (frames 20–21)

15.   the fact that it is coherent—its waves are all in phase   (frames 23–24)

CHAPTER TWENTY

Reflection, Refraction, and Dispersion

No Prerequisites. Chapters 10 and 11 suggested.

The phenomena which form the subject of this chapter—reflection, refraction, and dispersion of light—are among the most common of our everyday experience. The first two are so common, in fact, that we are seldom aware of their existence. The third is what gives us the beautiful colors of the rainbow.

When you have completed your study of this chapter you will be able to:

- compare the angles of incident and reflected light;

- differentiate between specular and diffuse reflection;

- determine relative size and position of the object and image in a plane mirror;

- differentiate between reflection and refraction;

- state the effect of speed of light in a material on refraction;

- compare the wavelengths of light in two materials where the speed of light differs;

- specify what the "normal" means in terms of reflection and refraction;

- specify the meaning of "total internal reflection" and state the conditions necessary for its occurrence;

- explain the effect of refraction on our perception of sunset;

- explain the cause of mirages in terms of the phenomena presented in this chapter;

- specify the relationship of dispersion to refraction;

- state the cause of rainbows;

- determine the angle of refraction, given the angle of incidence of light and the speed of light in each material (optional);

- determine the angle of refraction, given the angle of incidence of light and the index of refraction of each material (optional);

- determine the critical angle for total internal reflection between two substances, given the index of refraction of each (optional);

- calculate the speed of light in a substance given the index of refraction of the substance (optional).

REFLECTION

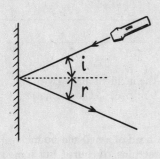

1.   If you shine a flashlight into a mirror in a room with a lot of dust in the air—perhaps you can find a mirror in a hayloft—you will be able to see the beam approach and reflect from the mirror. It will appear somewhat as in the figure at the left. The dashed line drawn perpendicular to the mirror at the point where the beam strikes it is called the "normal." Notice that the angle which the incoming (or "incident") beam makes with the normal is equal to the angle made by the reflected beam. This relationship between the angle of the incident beam (angle i) and the angle of the reflected beam (angle r) is called the law of reflection and is stated in formula form as:

$$i = r$$

(a)   According to the law of reflection, if a beam of light strikes a mirror at an angle of $35°$ with the normal, at what angle will the reflected beam

leave the mirror? _____

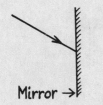

Mirror →

(b)   The figure at the left represents a beam of light striking a mirror. Sketch in the normal and the reflected beam.

- - - - - - - - - - - - - - - - - -

(a) $35°$ with the normal

(b)   The two angles marked in the figure are equal.

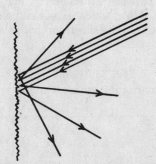

2.   When light is reflected from a plane (flat) mirror, the beam is reflected evenly as *specular reflection*. When light strikes a nonshiny surface such as the paper of this page, however, *diffuse reflection* occurs. The figure in this frame represents the surface of this page highly magnified. The light striking the surface

strikes tiny parts of the surface which are at different angles to the incoming rays. Since there is no regularity in the surface, there is no regularity in the reflected light—it bounces off in all directions. Diffuse reflection allows you to see this page (and most other objects) no matter at what angle they are with respect to you and to the incoming light.

(a)   What type of reflection allows you to see this book as you read?_____

(b)   What type of reflection allows you to see yourself in a mirror? _____

(c)   What is the formula for the law of reflection?_____

(d)   State the law of reflection in words. _____

_____

(a) diffuse; (b) specular; (c) angle i = angle r, or $i = r$; (d) the incident angle (with the normal) is equal to the reflected angle (again measured with respect to the normal)

3.   Figure 20.1 shows two rays of light coming from the head of a nail and bouncing from a plane mirror. Light is coming from the nail in all directions, of course, but some rays happen to strike the mirror and bounce off into the two eyes at left.

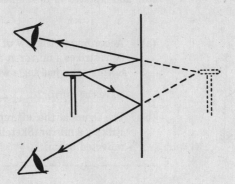

Figure 20.1

To the two eyes, the light entering them seems to have come from a single point behind the mirror. If the eyes are moved so that two more rays of light coming from the nail bounce into the eyes, these two rays will still seem to have come from exactly the same point as the two rays in the figure. In fact, every ray of light which leaves that point on the nail and strikes the mirror will rebound in a direction as if it came from that point behind the mirror.

(a)   What type of reflection are we discussing? _____

(b)   At what angle does the light from the nail leave the mirror if it strikes it at 80°

from the normal?_____

_____

(a) specular; (b) 80° from the normal

4.   We can analyze the light coming from any point on the nail in Figure 20.1. All of the reflected light will seem to have come from a point behind the mirror. In this way an *image* of the nail is formed by the mirror. In the figure, the distance from the nail to the mirror is the same as the distance from the image to the mirror and the image is the same size as the original nail.

These observations can be generalized to a law: The image of an object formed by a plane mirror is the same size as the object and is located behind the mirror at a distance equal to the object's distance in front of the mirror.

(a)   What type mirror forms an image the same distance behind it as the object is in front of it?

_____

(b)   The object shown at the left is near the mirror. Sketch the image.

– – – – – – – – – – – – – –

(a) plane; (b) The image you drew should be the same size as the object (the heart) and it should be an equal distance from the mirror.

REFRACTION

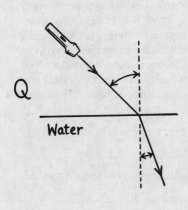

Q

Water

5.   Figure Q at the left shows the beam from a flashlight being shined into water. Again the normal is drawn in at the point where the beam enters the water. Compare the angles made by the incident light and the light after being bent—*refracted*—at the surface. The angle outside the water is greater.

Figure R shows a ray of light entering a flat plate of glass from the upper left. Notice that upon entering the glass it behaves just as did the beam entering water in the last figure. It bends so that its angle with the normal is less inside the glass than outside. Now let us focus on what happens when the light emerges from the glass. Again, a normal is drawn to the point where the light crosses the surface.

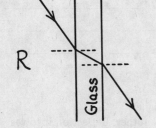

R

Glass

(a)   Compare the angles involved as the ray comes out of the glass. Which angle is largest?

_____

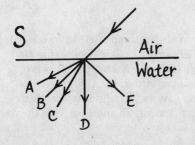

S

Air
Water

A
B
C
D
E

(b)   In Figure S a ray of light is shown about to enter water. Refer back to Figure Q. In which of the directions—A, B, C, D, or E—will the ray go after crossing the surface?

_____

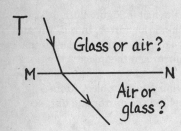

(c) Figure T shows the path taken by a ray of light as it crosses the boundary between glass and water. (Line MN represents the surface.) Is the glass above or below the line MN?_____

(d) Generalize from the three examples: When light crosses a boundary between air and water or air and glass, it bends so that the angle (with the normal) in the air is (greater/less) _____ than the angle in the other material.

_ _ _ _ _ _ _ _ _ _ _ _ _ _ _

(a) the angle outside the glass; (b) C; (c) above; (d) greater

6.   In order to see what causes light to bend when it crosses a surface, we will consider light as a wave rather than considering single rays of light. We have been considering the rays to be simply infinitesimally narrow beams of light. This is a convenient way to focus on the direction in which light is traveling. The figure here shows wavefronts (concentric circles) and rays coming from a point source of light. The rays are always perpendicular to the wavefronts.

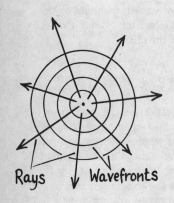

(a) Did we consider rays or waves to study reflection of light? _____

(b) Does a ray of light travel in the same direction as the wave travels?_____

(c) How does reflection differ from refraction? _____
_____
_____

_ _ _ _ _ _ _ _ _ _ _ _ _ _ _

(a) rays
(b) yes (they are just two different ways to consider the light)
(c) Refraction involves light going from one material into another. Reflection involves *bouncing from* the surface of another material.

7.   Figure 20.2 represents straight wavefronts moving in air from upper left to lower right and striking a surface (MN) of glass. The speed of light in glass is less than that in air, so the waves will slow down upon crossing the boundary. Look at what happens to wave CD. Although when it was at position AB, it was a straight wavefront, end C has already entered the glass, where it travels slower; this results in its covering less distance than end D. And the result of this is that the wavefront is no longer straight. The portion from C

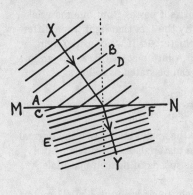

Figure 20.2

to the surface is now closer to parallel to the surface. Finally end D will enter the glass and the wavefront will be in a position such as EF. It is once again straight. But since the wave travels in a diarection perpendicular to the wavefront, it is now heading in a different direction than it was before entering the glass. The direction of travel of the light both before and after entering the glass is shown by the ray which stretches from X to Y.

(a)   How does the angle with the normal of the ray compare before and after entering the glass?

_____

(b)   What causes the light ray to change direction in this example? _____

- - - - - - - - - - - - - - - -

(a) larger outside the glass; (b) the change in speed of light

8.   Think of the waves of Figure 20.2 moving from lower right (in the glass) to upper left (in air). In this case side D has emerged first from the glass and has increased in speed, resulting in an increase in wavelength and a change in direction. Again, the law of refraction stated in an earlier frame is explained by the fact that light travels at different speeds in different materials. We might state the law of refraction qualitatively as follows: When light crosses a boundary between materials in which it travels at different speeds, it bends such that its angle with the normal is less in the material in which it travels more slowly.

(a)   Refer to Figure 20.2. Does light travel slower above or below the line MN?

_____

(b)   Is its angle with the normal less above or below line MN? _____

(c)   Light travels more quickly in a vacuum than in clear plastic. Which will have the greater angle with the normal? _____

- - - - - - - - - - - - - - - -

(a) below; (b) below; (c) the vacuum

Optional

9.   Thus far we have considered refraction only qualitatively. The quantitative statement of the law of refraction involves sines of the angles between the light rays and the normal. A table of the sines of angles is given in Appendix III.

(a)   What is the sine of 20°? (Use Appendix III.) _____

(b)   What angle has a sine of .575? _____

- - - - - - - - - - - - - - -

(a) .342; (b) 35° (Actually it is a little greater than 35°, but we will use the nearest whole degree.)

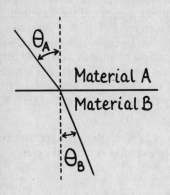

Figure 20.3

10.   Bending of light as it passes from one material to another is caused by the fact that light has a different speed in the two materials. If we consider light passing from material A into material B, as shown in Figure 20.3, the law can be stated as follows:

$$\frac{\text{sine } \theta_A}{\text{sine } \theta_B} = \frac{v_A}{v_B}$$

where θ_A and θ_B are the angles made with the normal by the rays of light in materials A and B respectively, and v_A and v_B are the speeds of light in materials A and B.

The speed of light in air is about $3 \cdot 10^8$ meters/second* and in water it is $2.25 \cdot 10^8$ m/s. Suppose a ray of light in air strikes a surface of water at an angle of 30° with the normal. At what angle with the normal does it travel in the water? (Refer to Appendix III for sines.)_____

- - - - - - - - - - - - - - -

22°.   Solution: Since it doesn't matter which is material A and which is B, we'll let water be A and air be B. "Sin" is the abbreviation for "sine."

$$\frac{\sin \theta_A}{\sin 30°} = \frac{2.25 \cdot 10^8 \text{ m/s}}{3.0 \cdot 10^8 \text{ m/s}}$$

$$\frac{\sin \theta_A}{0.5} = 0.75$$

$$\sin \theta_A = 0.375$$

$$\theta_A = 22°$$

11.   The quantitative law of refraction is usually stated differently from the equation in frame 10. To understand the way it is usually stated we must first define the index of refraction.

The index of refraction, n, of a material is the ratio of the speed of light in a vacuum, c, to the speed of light in that material. Or:

$$n_A = \frac{c}{v_A}.$$

In the example of frame 10, the index of refraction of water is $(3 \cdot 10^8)/(2.25 \cdot 10^8)$, or 1.33. From the equation defining the index we can see that:

$$v_A = \frac{c}{n_A}$$

---

*Appendix I is a discussion of powers of ten notation.

We can expand this to:

$$\frac{v_A}{v_B} = \frac{c/n_A}{c/n_B} = \frac{n_B}{n_A}$$

Thus we can rewrite the law of refraction as:

$$\frac{\sin \theta_A}{\sin \theta_B} = \frac{n_B}{n_A}$$

Or, as commonly stated:

$$n_A \cdot \sin \theta_A = n_B \cdot \sin \theta_B$$

Table 20.1 gives the indexes of refraction for a number of common substances.

Substance	Index of refraction
water	1.33
air	1.0003
crown glass	1.52
flint glass	1.66
diamond	2.42

Table 20.1

(a)   Refer to the table. What is the speed of light in crown glass? _____

(b)   Suppose material B in Figure 20.3 is diamond. A ray of light traveling in air strikes the diamond at an angle of 30° with the normal. What is the angle of the refracted ray? _____

(c)   Compare the answer to (b) above to the answer to the question in frame 10. In both cases the angle in air was 30°. Why does the ray refract (bend) more in diamond? _____

_____

- - - - - - - - - - - - - - -

(a) $1.97 \cdot 10^8$ m/s.   Solution: $v = c/n = (3 \cdot 10^8$ m/s$)/1.52$
(b)   12°.   Solution: $n_A \cdot \sin \theta_A = n_B \cdot \sin \theta_B$
$$1 \cdot \sin 30° = 2.42 \cdot \sin \theta_B$$
$$\frac{0.5}{2.42} = \sin \theta_B$$
$$0.206 = \sin \theta_B$$
(c)   The index of refraction is greater for diamond than for water. (Or, the speed of light is less in diamond than in water.)

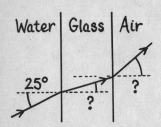

12.   In this frame we will work through another problem in detail. Suppose a ray of light is traveling in the water of an aquarium and strikes the glass wall at an angle of 25° (with the normal). We want to know first its angle in the glass and then in the air. See the figure at the left.

(a)   First solve for the angle in the glass, using the equation for the law of refraction, and assuming crown glass. _____

(b)   Once you know the angle made with the normal in the glass, you know the angle at which the ray strikes the glass-air surface on the right—the same angle. Now

what is the angle in air at the right? _____

(a)   22°. Solution: Letting the glass be material A:

$$1.52 \cdot \sin \theta_A = 1.33 \cdot \sin 25°$$

$$\sin \theta_A = \frac{1.33}{1.52} \cdot 0.423$$

$$\sin \theta_A = 0.370$$

$$\theta_A = 22°$$

(b)   34°. Solution: Letting A be the air this time:

$$1.0 \cdot \sin \theta_A = 1.52 \cdot \sin 22°$$
$$\sin \theta_A = 1.52 \cdot 0.370$$

Note that the value of the sine from above is used, rather than the 0.375 given in the table. The angle is not really 22°, but a little less. If you use 0.375 the difference in your answer and mine will be very little, though.

$$\sin \theta_A = 0.564$$
$$\theta_A = 34° \text{ (or a little more)}$$

TOTAL INTERNAL REFLECTION

13.   In the figure below, part A shows a ray of light passing from glass into air. It makes an angle of 30° with the normal in glass and about 50° in air. In figure B, the corresponding angles are 35° and 60°. (If you studied the quantitative refraction section, you can

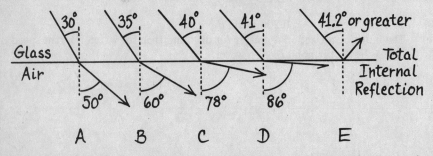

check these out, using an index of refraction of 1.52.) In figure C, the angles are 40° and 78°. In D they are 41° and 86°.

The angle in the air is approaching 90°. Once it reaches this point, no light is refracted, but all of it is reflected back into the material from which it hit the surface (figure E). This phenomenon is called *total internal reflection*. It occurs when two conditions are met:

(1)   The light is moving toward the surface of a material in which it travels faster; and

(2)   It hits the surface at an angle (with the normal) greater than some critical angle.

The critical angle for light moving in the glass above toward an air surface is 41.2 degrees.

(a)   What happens if light traveling in this glass comes to the air-glass surface at an angle less than 41.2°?_____

(b)   What happens if it hits the surface at an angle greater than 41.2°?_____

_____

(c)   What happens if a light beam is moving in air towards a glass surface at an angle of 60°?_____

— — — — — — — — — — — — — — —

(a) it is refracted through the surface; (b) it is totally internally reflected; (c) it is refracted into the glass (Condition 1 is not met—the light is not inside, "internal to," the glass.)

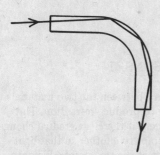

14.   Total internal reflection is used in a practical application in light pipes. The figure at the left represents a solid cylindrical filament of glass (or plastic). Light has entered the left end, and at each encounter with a surface, its angle with the normal is greater than the critical angle at which it can emerge from the glass (or plastic). This results in the light being carried on down the tube to the other end.

Light pipes are used routinely today in medicine and industry. They have been proposed as a replacement for wires to carry telephone and cable television signals. Instead of electricity in wires, light would carry the signal in extremely fine light pipes. The advantages* are sufficient to allow one to predict that light pipes for communication will become commonplace in the future.

(a)   How does a light pipe differ physically from a hollow plastic tube?_____

_____

(b)   Why doesn't the light leave the light pipe?_____

_____

— — — — — — — — — — — — — —

_____

*Two advantages: there would not be a loss of power and resultant heating of the light pipe as there is for wires, and a single light signal is capable of carrying a very great number of communications at the same time.

(a) it is not hollow, but solid; (b) it strikes the surface at greater than the critical angle

Optional. Prerequisite: frames 10–11

15.   The critical angle for total internal reflection is the angle that would make the light refract at an angle of 90° with the normal. If we let material A be the "internal" material, we can write:

$$n_A \cdot \sin \theta_A = n_B \cdot \sin 90°$$

Since the sine of 90° is 1.00, we can write:

$$\sin \theta_{crit} = \frac{n_B}{n_A}$$

Where the external material is air, we have:

$$\sin \theta_{crit} = \frac{1}{n_A}$$

For example: Consider the critical angle for a water-air boundary.

(a)   In which material would the phenomenon of total internal reflection occur?

Why? _____

(b)   What is the critical angle? _____

– – – – – – – – – – – – – – – –

(a)   water, because light travels slower in it than in air
(b)   49°. Solution: $\sin \theta_{crit} = \frac{1}{1.333}$
$$\sin \theta_{crit} = 0.75$$

ATMOSPHERIC REFRACTION: SUNSET

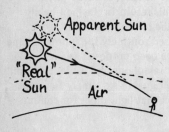

16.   A definite boundary between the two materials is not actually necessary to produce refraction. The phenomenon will also accur if there is a gradual change in a material from one point to another so that light travels at different speeds at the two points. We have been considering the speed of light in air to be the same as the speed of light in a vacuum, but actually light travels slightly slower in air.

In the figure at the left light from the setting sun is shown entering the atmosphere and bending down to an observer. Notice that the light coming to the person seems to be coming to him from higher in the sky than the actual position of the sun. He sees the sun as being higher in the sky than it actually is! In fact, we can see the sun after it has set (if we define "set" based upon a straight line from the sun to us).

(a)   What phenomenon enables us to see the sun after it has set? _____

(b)   What causes the phenomenon here? _____

– – – – – – – – – – – – – – –

(a) refraction; (b) light travels slower in air than in a vacuum

17. The slight difference in speed between a vacuum and air means that atmospheric refraction is not very great until the angle at which the light strikes the atmosphere is great. This is noticeable on a very clear evening just before the sun seems to be setting on a low horizon (the ocean, perhaps). In such an instance, the light from the "bottom" of the sun is bent more than the light from the "top," and the sun appears to be flattened. The next time you see a really clear sunset, look for this. Although refraction by the atmosphere is not normally great, astronomers who are plotting star positions must take it into account, for in precise measurements the refraction is enough to make a difference.

When the setting sun appears to be flattened, are light rays from the top or bottom

more strongly refracted? _____

— — — — — — — — — — — — — —

bottom

MIRAGES

18. Light travels slightly faster in hot air than in cooler air. In the figure below, the air near the hot road is hotter than the air farther above. This results in light from the sky being refracted back up from the road. The ray of light in the figure bends up to the eye of the driver. To the driver, it appears that the road is *reflecting* blue light from the sky, and he is likely to think that the road ahead appears to be wet. Look for this the next time you ride on a hot asphalt road on a still summer day (with no wind to stir the air).

This is the same phenomenon that results in thirsty desert travelers seeing water-holes ahead, only to find them disappearing when they are approached. In this case, imagination also plays a bit role.

What phenomenon causes mirages? _____

— — — — — — — — — — — — — —

refraction

DISPERSION

19. In what has been said so far, we have ignored the fact that all frequencies of light do not travel at the same speed in all materials. It is true that all frequencies travel at the same speed in a vacuum, but in all other materials the speed of the lower frequencies is greater than the speed of higher frequencies. Since the visible spectrum goes from red at the lowest frequency through orange, yellow, green, blue, and finally violet at the greatest frequency, we would expect the various colors to be bent at different angles when they are refracted at a boundary. Low frequencies have the greater speed in glass, so we would expect violet to bend most when light enters glass.* And indeed, this separation of colors

---

*Since violet travels more slowly in glass, it slows down more upon entering. It is this slowing down which causes the bending.

occurs at every refraction of light. It is not observed as light passes through a window for two reasons. First, the separation by thin windows is very slight. Second, when the light emerges, the violet that was separated from the red of one ray emerges with the red of another ray and no one is the wiser.

The figure at the left shows light going through a prism. In this case, the sides of the glass are not parallel, so that the effect produced by the first side is enhanced by the second side. This separation of light into colors by refraction is called *dispersion*.

(a) Is dispersion more apparent when light passes through a flat sheet of glass or through a prism?

_____

(b) Which color light is refracted most by glass? Why? _____

_____

(c) Is dispersion a result of refraction or reflection? _____

— — — — — — — — — — — — — — — —

(a) prism; (b) violet, because it travels slower in glass than do the other colors; (c) refraction

THE RAINBOW

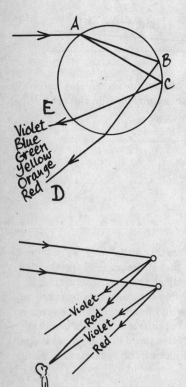

20. The figure at the left shows a ray of white light striking a spherical drop of water at point A. The light is dispersed slightly, with the various colors hitting the other side of the drop between points B and C. At these locations some of the light is refracted out of the drop, but we are interested in the part of the light that is reflected back toward the region from D to E. Here again some light is refracted out of the drop and some is reflected. It is the light which emerges at the surface between D and E that results in rainbows.

The second figure in this frame shows two drops being hit by sunlight coming from the left. Light reaching the person's eye from the higher drop is red, while light from the lower drop is violet. Light from raindrops in between these sends light of the other colors of the spectrum to the rainbow-watcher. Since the angles necessary to see each of the colors occur not only along a vertical line, but to the sides as well, a curved bow is seen.

(a) What phenomenon of this chapter causes

rainbows? _____

(b)   Why must the sun be shining for you to see a rainbow? _____

(c)   If you are to see a rainbow at sunrise, what direction must you look? _____

— — — — — — — — — — — — — — —

(a) dispersion; (b) to provide the light (coming from a single direction); (c) west, since the sun rises in the east and the rainbow must be opposite to it, and thus arches across the west

SELF-TEST

The questions below will test your understanding of this chapter. Use a separate sheet of paper for your diagrams or calculations. Compare your answers with the answers provided following the test.

1.   What is the law of reflection? _____

      _____

2.   What is the difference between diffuse and specular reflection? _____

      _____

      _____

3.   What is the relationship between the sizes and distances-to-the-mirror of the object

      and image formed by a plane mirror? _____

      _____

4.   Suppose you find a figure such as the one at the right
      in a textbook. It represents light passing through a
      boundary between glass and air. The author, however,
      has forgotten to tell you on which side of the surface
      is the glass. Is the glass above or below the surface?

      _____

5.   Suppose light travels from air into glass (where it
      travels slower). Is its wavelength greater in the air or

      in the glass? _____

6.   What conditions are necessary in order for the phenomenon of total internal reflec-

      tion to occur? _____

      _____

7.   What phenomenon causes us to be able to see the sun after it has set below the

      horizon? _____

8.   Are mirages caused by reflection or refraction? _____

9.   Blue light travels slower than yellow light in glass. Which of the two colors will be bent more by a glass prism? _____

10. Although both reflection and refraction must be used in the explanation of the rainbow, which causes the dispersion? _____

11. *Optional.* Does light travel faster in crown glass or in flint glass? (Use Table 20.1.)

_____

12. *Optional.* Suppose light traveling in water strikes the boundary with air at an angle of 25° (with the normal). At what angle does it travel in the air? _____

13. *Optional.* Is the critical angle for diamond greater or less than that for glass? (Use Table 20.1.) _____

14. *Optional.* What is the critical angle for a diamond-air boundary? _____

Answers to Self-Test

If your answers do not agree with those given below, review the frames indicated in parentheses before you go on to the next chapter.

1.   The angle of incidence is equal to the angle of reflection.   (frame 1)

2.   Specular reflection is the regular, orderly reflection such as occurs in the case of mirrors. Diffuse reflection is reflection-in-all-directions such as occurs from a non-shiny surface.   (frame 2)

3.   The object and image are the same size and are the same distance from the mirror. (frame 3)

4.   above (The ray is closer to the normal—not shown—above the boundary, so we know that the light travels slower there.)   (frame 7)

5.   The wavelength is greater in air because light has a greater speed there.   (frame 7)

6.   The light must be coming toward the material in which it moves faster, and it must be incident upon the surface at an angle greater than the critical angle. (frame 13)

7.   refraction   (frame 16)

8.   refraction   (frame 18)

9.   The blue light will bend more.   (frame 19)

10. refraction   (frame 20)

11. crown glass (Since crown glass has the lesser index of refraction, it follows from the definition of the index that light travels faster in that material.)   (frame 11)

12.   34°. Solution: calling air material A:

$$1.000 \cdot \sin \theta_A = 1.33 \cdot \sin 25°$$
$$\sin \theta_A = 1.33 \cdot .423$$
$$= .563 \quad \text{(frames 11–12)}$$

13.   greater (Since the sine of the critical angle is equal to the reciprocal of the index of refraction, a greater index of refraction results in a smaller critical angle.) (frame 15)

14.   24°. Solution: $\sin \theta_{crit} = \frac{1}{2.24} = .413$ (frame 15)

CHAPTER TWENTY-ONE

Lenses and Instruments

Prerequisite: Chapter 20

We have seen how the phenomenon of specular reflection is effectively used in mirrors. The phenomenon of refraction is widely used in lenses. In this chapter we will explore the varied applications of refraction in lenses.

After studying this chapter, you will be able to:

- differentiate between converging and diverging lenses;

- define the focal length and focal point of a lens;

- differentiate between real and virtual images;

- relate image position to focal length and object distance for convering and diverging lenses;

- determine magnification produced by a given set of conditions;

- explain the function of component parts of a camera—diaphragm, film, and lens;

- compare the operation of a camera to the operation of the human eye;

- indicate the type of lens needed to correct the two common eye defects;

- given a diagram, explain how a projector operates;

- indicate the object for each lens in a microscope;

- determine the total magnification of a two-lens microscope given relative distances of objects and images;

- indicate the primary difference between a microscope and a telescope;

- specify the two major functions of a telescope and what parts of the telescope contribute to each;

- calculate the magnification of a telescope given the focal lengths of the two lenses;

- use the lensmaker's formula to calculate the focal length of a lens given the radii of curvature of the sides and the index of refraction (optional);

- calculate the position of an image, given focal length and object position for both converging and diverging lenses (optional).

LENSES

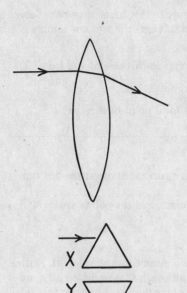

1.   We saw in the last chapter (frame 19) that light going through a prism changes direction (as well as separating into colors). This bending occurs because the two sides of the glass are not parallel. The figure at the left illustrates a glass lens with one ray of light going through it. If you draw normals to the surfaces of the glass where the ray crosses them, you can verify that the law of refraction is obeyed in each case.

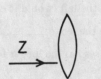

(a)   In which of the cases shown at left, X or Y, will the ray of light be bent *upward* by the prism?

_____

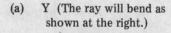

(b)   What will happen to the ray of light about to strike the lens in figure Z? _____

_____

– – – – – – – – – – – – – – – –

(a)   Y (The ray will bend as shown at the right.)

(b)   It will bend upward.

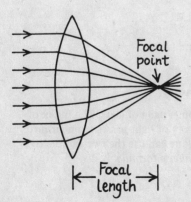

Focal point

Focal length

2.   The figure in this frame shows a great number of rays passing through a lens. After they pass through, they come together, or converge. Lenses such as this—thicker at the center than at the edges—are called converging lenses. They are three-dimensional, appearing somewhat like that shape obtained by inverting one dinner plate on top of another. In fact, if the lens is made correctly (as is the one shown), they converge so that they cross at a single point. In this figure the rays are parallel to one another and to the axis of the lens as they enter at the left. The point at which parallel rays converge after passing through a lens is

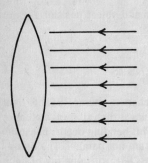

called the *focal point* of the lens. The distance from the lens to the focal point is called the *focal length* of the lens.

The second figure in this frame shows the same lens as before, but in this case the rays are coming from the right.

(a)   Will they converge in this case or will they diverge (spread out)? _____

(b)   Where will the focal point be located?

_____

- - - - - - - - - - - - - - -

(a)   converge (Those at the top will be bent downward again and those at the bottom will bend upward.)

(b)   to the left of the lens (All lenses have two focal points, and the points are located at equal distances on each side of the lens.)

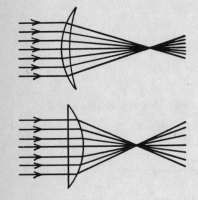

3.   Here we see two more converging lenses. Notice that the two lenses, although shaped differently, are both thicker at the center than at the edges. It is this feature which causes them to be converging lenses.

(a)   If the light had struck the upper lens from the right instead of from the left, would it converge after passing through the lens? _____

(b)   Which of the two lenses has the longer focal length? _____

- - - - - - - - - - - - - - -

(a) yes; (b) the upper one

Optional. Prerequisite: Chapter 20, frames 10–11

THE LENSMAKER'S FORMULA

4.   Two factors influence the focal length of a lens:

the curvature of the lens surfaces

the index of refraction of the glass (or plastic)

The curvature is what makes a lens act differently from a pane of glass. The index of refraction determines the amount of bending when a ray of light passes from air into the lens (and then back out). The equation by which one can use the two factors to calculate the focal length of a lens is called the *lensmaker's formula:*

$$\frac{1}{f} = (n - 1)\left(\frac{1}{r_1} + \frac{1}{r_2}\right)$$

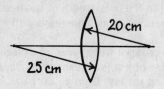

where f is the focal length of the lens, and r_1 and r_2 are the radii of curvature of the two lens surfaces.

The figure at the left shows a lens with radii of curvature of 20 cm and 25 cm and with glass of index of refraction 1.52. Use the formula to calculate its

focal length. _____

$f = 21.4$ cm. Solution: $\frac{1}{f} = (1.52 - 1) \left(\frac{1}{20} + \frac{1}{25} \right)$

$= (0.52)(0.05 - 0.04)$

$= 0.0468$

$f = \frac{1}{0.0468}$

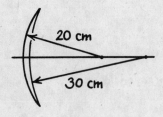

5.   In the lens you just used, all of the values were positive. If one of the surfaces of the glass were curved in the other direction, its radius would be negative. For example, the right surface of the lens shown here is backwards from the "standard" above. Consider its radius to be *negative* 30 cm. What is the focal length

of this lens? _____

$f = 120$ cm. Solution: $\frac{1}{f} = (1.52 - 1) \left(\frac{1}{20} + \frac{1}{-30} \right)$

$= (0.52)(0.05 - 0.033)$

$= 0.0833$

IMAGES

6.   In the figure below, two sets of parallel rays are seen coming toward a lens from the left. They could represent rays from the top and the bottom of a very distant object. (If an object is distant enough, the light rays from any point on it are essentially parallel.) Note that the rays from each point on the object again converge to a point after passing

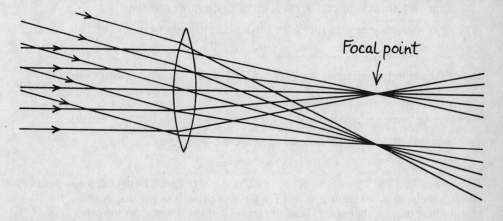

Focal point

through the lens, and that this point is at a distance equal to the focal length of the lens. One set of rays focuses *below* the focal point, however, because its incoming rays were not parallel to the axis of the lens.

If the top of the distant object is a red light bulb, a red spot will appear on a piece of paper held where the rays converge. If the bottom of the object is a green light bulb, a green spot will appear on the paper above the red spot. And the rays from points on the object between the top and bottom will appear on the paper also. In fact, an image of the object will be seen on the paper.

(a) Will the positions of the red and green spots on the paper be inverted from the positions of the lights? _____

(b) Is the distance between the red and green spots on the paper greater or less than the distance between the red and green light bulbs? _____

– – – – – – – – – – – – – – – –

(a) yes; (b) less

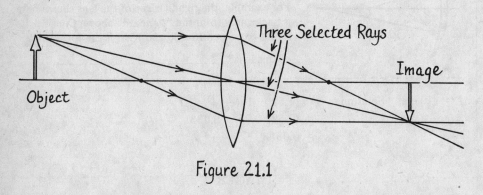

Figure 21.1

7.   In Figure 21.1 a nearby object is shown. In this case, the image will not appear at the focal plane (the vertical plane through the focal point), but will be more distant from the lens. With a nearby object, the rays are spreading out (diverging) when they reach the lens and the lens is unable to converge them as close to it as it can parallel rays. As in the case of the very distant object, the image is upside down, or inverted.

(a) If an object is moved farther from a lens, what happens to its image? _____

_____

(b) If the object is extremely far away (at "infinity"), where is the image located?

_____

– – – – – – – – – – – – – – – –

(a) it gets closer to the lens; (b) on the focal plane

8.   We call the images we have been considering *real* images. The light actually passes through the position of the image, and if a piece of paper is put at that position, the image will appear on the paper. We saw in frame 3 of the last chapter the other type of

image, the *virtual* image. Recall that the image formed by a flat mirror is located behind the mirror and that therefore no light is actually at the position of the image. The light bouncing from the mirror only appears to come from the image. Such images are called virtual images.

(a)   If a piece of paper were put at the position of the image of a plane mirror, would the image appear on the paper?_____

(b)   When you watch a movie on a screen, are you seeing a real or a virtual image?

_____

(c)   When you look at yourself in a mirror, is the image you see real or virtual?

_____

– – – – – – – – – – – – – – – –

(a) no; (b) real, because light actually comes from the position of the image on the screen; (c) virtual

MAGNIFICATION

9.   In Figure 21.1 the image was closer to the lens than was the object, but the object was larger. Now examine Figure 21.2. Here the object is closer to the lens, but the image is larger than the object. In general, the relationship between the object and image size, and the object and image distance is as follows:

$$\frac{I}{O} = \frac{q}{p}$$

where
I = image size
O = object size
q = image distance from the lens
p = object distance from the lens

Magnification is generally defined as being the image size divided by the object size, but it is also equal to the image distance divided by the object distance.

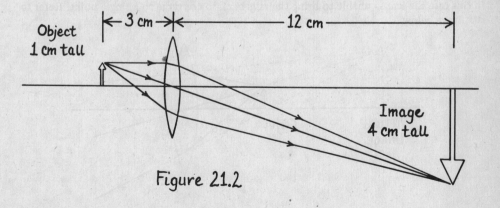

Figure 21.2

(a)   Use the distances indicated in Figure 21.2 to calculate the magnification of the lens as it is being used there. _____

(b)   If the object were placed twice as far from the lens, would the magnification be the same? _____

- - - - - - - - - - - - - - -

(a) 4; (b) no (Recall that as the object moves farther away, the real image moves closer.)

Optional   IMAGE CALCULATIONS

10.   The relationship between the focal length f of a lens, the object distance p, and the image distance q, is:

$$\frac{1}{f} = \frac{1}{p} + \frac{1}{q}$$

Again, all quantities are positive when a converging lens is used and the object and image are on opposite sides of the lens (which is the only case we have considered so far).

Now suppose you have a lens with a focal length of 12 cm and you locate an object 20 cm from it. Where will the image be? _____

- - - - - - - - - - - - - - -

30 cm from it (on the oppoiste side).   Solution: $\frac{1}{12} = \frac{1}{20} + \frac{1}{q}$

$0.0833 = 0.05 + \frac{1}{q}$

$0.0333 = \frac{1}{q}$

VIRTUAL IMAGES: THE MAGNIFIER

11.   When an object is very distant from the lens, we saw that the image is located at the focal point, or focal plane. As the object moves closer, the image moves farther from the lens. Finally, when the object is put at the focal point, the image is located very far away (at "infinity," actually). What happens when the object is placed inside the focal point? In this case the lens is unable to bring the rays back to converge to a single point. Refer to the figure below.

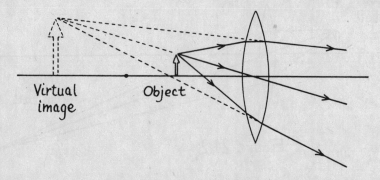

Virtual image   Object

You will see that the rays from the top of the object, after coming through the lens, *seem* to be coming from a point farther back on the left. In fact, an image of the object is formed there. The rays of light do not actually go through the image, so the image is a virtual one. If a piece of paper were put at the location of this virtual image, an image would not appear on the paper. We can say that an image is there, however, because if the rays going toward the right enter an eye (or a camera) it appears that the object is located where the image shows in the diagram.

(a)   Which is larger in the case we are discussing, the object or the image?_____

(b)   Which is farther from the lens?_____

(c)   In what way is this image similar to the image in a plane mirror?_____

_____

— — — — — — — — — — — — — — — —

(a) the image; (b) the image; (c) both are virtual images

12.   Refer to Figure 21.3. Notice that the virtual image is larger than the object again. (This will always be the case when the object is inside the focal point of a converging lens.) Imagine that you are looking at the situation from the right (so that the rays pass into your eyes). A lens used in this manner is often called a magnifying glass, or a magnifier. The relationship between sizes and distances holds for this case just as for the real images studied before.

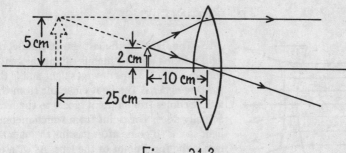

Figure 21.3

(a)   Calculate the magnification in Figure 21.3 using the ratio of image size to object size. _____

(b)   Calculate the magnification using the ratio of distances—image distance to object distance. _____

— — — — — — — — — — — — — — — —

(a) 2.5; (b) 2.5

Optional. Prerequisite: frame 10

CALCULATIONS OF VIRTUAL IMAGE DISTANCES

13.   The equation relating focal length, object distance, and image distance, $1/f = 1/p + 1/q$, also holds in the case of virtual images. A negative sign must appear with the image distance, however, since it is on the opposite side of the lens from what it is in the "standard" situation.

(a)   Calculate the focal length of the lens in Figure 21.3 using the object and image distances given there. _____

(b)   Suppose an object is located at a distance of 6 cm from a lens with a focal length of 10 cm. Where will the image be located? _____

(a)   16.7 cm. Solution: $\frac{1}{f} = \frac{1}{10} + \frac{1}{-25}$ (Note the minus sign.)
$$= 0.10 - 0.04$$
$$= 0.06$$

(b)   −15 cm (the minus sign indicates that the image is on the same side of the lens as the object).   Solution: $\frac{1}{10} = \frac{1}{6} + \frac{1}{q}$
$$0.1 = 0.167 + \tfrac{1}{q}$$
$$\tfrac{1}{q} = -.067$$

DIVERGING LENSES

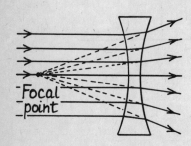

14.   The figure at the left shows parallel rays of light striking a lens which is thinner at center than at the edge. Such a lens diverges light and is thus called a *diverging lens.* The light emerging from the lens acts as if it comes from a point closer to the lens than it actually does. The point from which incoming parallel light seems to come after passing through a diverging lens is the focal point of the lens. As with converging lenses, the distance from the focal point to the lens is the focal length.

Figure 21.4 shows the formation of an image by a diverging lens. The image formed by such a lens is always smaller than the object. (Figure 21.4 is shown at the top of page 251.)

(a)   Is a real or a virtual image formed by a diverging lens? _____

(b)   What type lens can be used as a magnifier? _____

(a) virtual; (b) converging

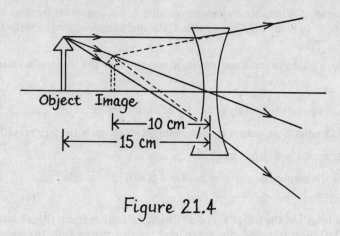

Object   Image

|←——10 cm——→|

|←———— 15 cm ————→|

Figure 21.4

Optional. Prerequisites: frame 10

DIVERGING CALCULATIONS

15.   Image and object distances for diverging lenses are calculated as for converging lenses. But now we have another sign change, since the focal length of a diverging lens is negative. Use the given object and image distances to calculate the focal length of

the lens of Figure 21.4. _____

−30 cm. Solution: $\frac{1}{f} = \frac{1}{15} + \frac{1}{-10}$ (image on left, thus minus sign)

$= -0.333$

THE CAMERA

16.   The basic camera is perhaps the simplest possible use of a lens. The figure below is a diagram of a box camera. An image of the distant object is formed on the film.

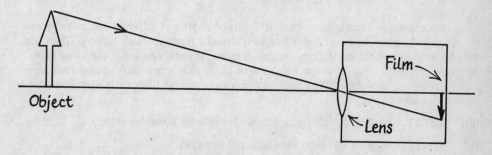

Film

Object

Lens

    You will recall that when the object distance is changed, the image distance is likewise changed. The simplest cameras do nothing about this problem. Their lens is set so that a fairly distant object is in good focus. A better camera has an adjustable lens-to-film distance so that it can be focused for objects at various distances. In addition, a good camera contains an adjustable diaphragm opening which can be set to allow different amounts of light to reach the film. In effect, the diaphragm adjusts to make the lens larger or smaller.

(a) To take a photo of a distant object, what should be the lens-to-film distance?

_____

(b) Is the image on the film right side up or upside down? _____

(c) When an adjustable camera is used to take a close-up, is the lens moved closer to or farther from the film? _____

(d) How is the diaphragm changed to take a photo in dim light? _____

— — — — — — — — — — — — — —

(a) the focal length of the lens; (b) upside down; (c) farther from (Recall from frame 7 that when the object is moved closer, the image moves farther from the lens.); (d) it is opened, to allow more light in

THE EYE

17. The figure here illustrates the human eye. It operates very similarly to the camera (or vice versa, since the eye came first). Light passes through the cornea and lens to focus an image on the retina. The iris controls the amount of light entering the eye.

What part of the camera corresponds to each eye part below?

(a) iris _____

(b) retina _____

(c) lens and cornea _____

— — — — — — — — — — — — — —

(a) diaphragm; (b) film; (c) lens

18. The adjustable camera allows you to focus on objects at different distances by moving the lens. The eye, however, has a lens which can be made to change shape. When the muscles around it contract, the lens assumes a more curved shape and is therefore more converging. Such shortening of the focal length of the lens is necessary in order to bring a nearby object into focus. This is because it must reconverge the greatly diverging rays from the nearby object.

(a) Would the eye lens be thinner or thicker for viewing distant objects? _____

(b) What type image is formed on the retina of the eye? _____

— — — — — — — — — — — — — —

(a) thinner; (b) real

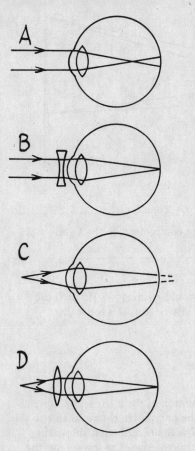

19.   The most common vision defect for young people is nearsightedness (myopia). This is caused either by the eyeball being too long or the lens being too powerful (too converging). The result is that the image is brought to a focus in front of the retina. (See figure A.) Such a defect is corrected by placing a diverging lens in front of the eye in order to spread the light and make it focus farther from the lens. This is shown in figure B.

Farsightedness (hyperopia) is the condition in which the image is (or would be) formed behind the retina. This is caused either by an eye which is too short or by a lens which is not converging enough. This condition is most common in middle and later ages, when the lens becomes somewhat stiff and the muscles are unable to squeeze it down into greater curvature. Figures C and D illustrate this defect and its correction.

(a)   Suppose the lens of figure B were used for the eye condition in C. Where (approximately) would an image be formed? _____

_____

(b)   What type lens is used to correct nearsightedness? _____

(c)   If the fault of the eye is that its lens is too thin, what type eyeglass lens corrects this?

_____

- - - - - - - - - - - - - - - -

(a) farther behind the retina; (b) diverging; (c) converging, since it is thicker at the center than at the edges

THE PROJECTOR

20.   The projector is, in some ways, the reverse of the camera. The figure at the top of page 254 shows a frame of movie film (or a slide) being illuminated by a light and the image being projected onto a distant screen. The screen is at a great distance compared to the focal length of the lens, while the film is placed very close to the focal point. A lens may be placed between the projector lamp and the film to concentrate light on the film, or a curved mirror may be placed behind the lamp to accomplish the same effect. In addition, projectors include a shield to block the light while changing slides on a slide projector, or while the movie film is going from one frame on the film to another. This shield would be located just to the left of the film in the figure.

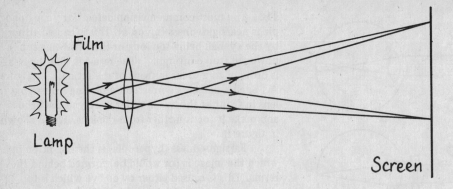

(a)   What type lens is used to project the image onto the screen? _____

(b)   If the image on the screen were out of focus, what components in the figure could

be adjusted to bring it into focus? _____

– – – – – – – – – – – – – – – – –

(a) converging; (b) the distance between film and lens would be changed (In fact, the distance between the lens and screen could be changed instead, but this is usually impractical.)

THE MICROSCOPE

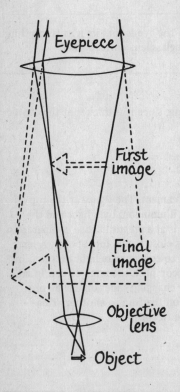

21.   Until the invention of the microscope, the simple magnifier was the only instrument available to see very tiny objects. The figure at the left shows the optical system of a microscope. The lower lens, the *objective*, has a very short focal length, and the object to be viewed is placed just beyond its focal point. This means that the image formed by the lens will be relatively far from the lens, and that it will be much larger than the object (remember from frame 9 the relationship between sizes and distances from the lens).

The upper lens, the *eyepiece*, uses the image formed by the objective as its object. From the figure you can see that the final image is not only much enlarged, but is inverted in orientation from the original object.

The magnification produced by a microscope is the product of the magnification of each of the lenses.

(a)   If the image formed by the objective is 25 times as far from the lens as is the object, what magnification does this lens produce? _____

(b) If the final image is 10 times as far from the eyepiece as is *its* object (the image formed by the objective), then what is the magnification of the microscope as a whole? _____

(a) 25 (see frame 9); (b) 250 (10 · 25)

THE TELESCOPE

22. The major difference between the function of a telescope and a microscope is that the object being viewed with a telescope is at a great distance while for a microscope it is close. The figure below shows the optical system of a refracting telescope.

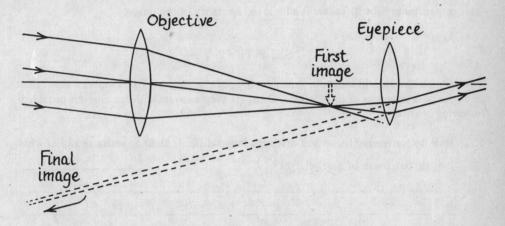

Since the object is very distant, the incoming rays are parallel. The first lens (the objective again) produces an image at its focal point. Then, just as with the microscope, the eyepiece is used as a magnifier to view the image.

In normal use of a telescope, the eyepiece is located at such a distance from the image formed by the objective that the final image is very far away—at infinity. It is thus meaningless to compare object and image distances to calculate magnification. What is important when one looks through a telescope at the moon or a planet is the relative angle taken up by the rays as they enter the eye. It turns out that magnification, when defined in this way, is equal to the focal length of the objective lens divided by the focal length of the eyepiece. A telescope has a single large objective lens, so in order to change the telescope's magnifying power, the viewer changes eyepieces. To get a high magnification, does one use an eyepiece with a long or a short focal length? _____

----

short (because magnification = $f_{\text{objective}} \div f_{\text{eyepiece}}$)

23. Since eyepieces of extremely short focal length can be made, there is theoretically no limit to how powerful a telescope can be made. Practically, however, telescopes do have a limit to power. Even on the clearest night, the atmosphere between us and the stars blurs the image if too much magnification is used. Magnification beyond a certain

point is worthless—it just magnifies the fuzziness of the image. A telescope is also limited by the wavelength of the light, which will be discussed in the next chapter.

Telescopes are used not just for magnification, but more importantly to gather as much light as possible from dim stellar objects so that we can see them. For this reason, telescopes are constructed with large objective lenses. The largest objective lens in use on a refracting telescope is 40 inches in diameter (at the Yerkes Observatory).

(a)   What is the advantage of a telescope with a large objective lens? _____

_____

(b)   Why is greater magnification not always desirable? _____

_____

– – – – – – – – – – – – – – – –

(a) it gathers more light; (b) there is a limit to the clarity of the image

SELF-TEST

The questions below will test your understanding of this chapter. Use a separate sheet of paper for your diagrams or calculations. Compare your answers with the answers provided following the test.

1.   How do converging lenses and diverging lenses differ in their appearance and in what they do to a beam of parallel light? _____

_____

_____

2.   How does the focal point for a converging lens differ from the focal point for a diverging lens? _____

_____

_____

3.   Suppose a converging lens forms an image of a very distant object. Where is the image located? _____

4.   As a distant object moves closer to a lens, what happens to the position of the image? _____

_____

5.   What is the relationship between magnification and image and object distances?

_____

6.   An object is 23 cm from a lens, and the magnification produced by the lens is 3. Where is the image located? _____

7.   What is the difference between real and virtual images? _____

_____

_____

8.   Compare the sizes of the object and image when a diverging lens is being used.

_____

9.   How does the eye accommodate (focus on objects at different distances)?

_____

10.   What are the two most common vision defects? What type lens is used to correct

each? _____

11.   What forms the object which is viewed by the eyepiece of a microscope? _____

_____

12.   Why is a large objective lens desirable for a telescope? _____

_____

13.   What is the magnification of a telescope with an objective of focal length 1.0 meter

(100 cm) when an eyepiece of focal length 2.0 cm is being used? _____

14.   If the objective of a microscope magnifies the object 20 times and the eyepiece pro-

duces a magnification of 15, what is the magnification of the microscope? _____

15.   *Optional.* The lens diagrammed below is made of glass with an index of refraction

of 1.6. What is its focal length? _____

16.   *Optional.* If an object is placed 25 cm in front of the lens of the last question, where

will the image be located? _____

17. *Optional.* What is the focal length of the lens shown below if the index of refraction of the glass is 1.52? _____

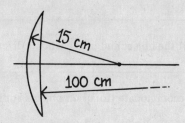

18. *Optional.* If an object is located 42 cm to the left of the lens of the last question, where will the image be located? _____

19. *Optional.* Suppose one wishes to use a lens of focal length 20 cm as a simple magnifier and have the image located 25 cm from the lens. Where must the object be placed? _____

20. *Optional.* A lens has a focal length of minus 20 cm. What type lens is it? _____
_____

Answers to Self-Test

If your answers do not agree with those given below, review the frames indicated in parentheses before you go on to the next chapter.

1. A converging lens is thicker at the center than at the edges while a diverging lens is the opposite. A converging lens brings parallel rays of light to a single point. A diverging lens spreads the rays so that they appear to have come from a single point behind the lens.   (frames 2, 14)

2. The focal point of a converging lens is that point to which rays which are parallel to one another and to the axis of the lens will converge after passing through the lens. For a diverging lens, the focal point is the point from which such rays seem to be coming.   (frames 2, 14)

3. at the focal point (or, to be more precise, on the focal plane)   (frame 6)

4. As the object approaches, the image recedes from the lens. When the object reaches the focal point, the image is at infinity.   (frame 7)

5. Magnification is equal to image distance divided by object distance.   (frame 9)

6. 69 cm from the lens (magnification = q/p, so $q = p \cdot$ magnification = $23 \cdot 3$)   (frame 9)

7. A real image is one which is actually formed by light passing through the image position. In the case of a virtual image, the light appears to come from the image, but it does not actually converge at the image.   (frame 8)

8. The image is always smaller than the object.   (frame 14)

9. The lens of the eye changes shape.   (frame 18)

10. myopia (nearsightedness)—corrected by a diverging lens
    hyperopia (farsightedness)—corrected by a converging lens   (frame 19)

11. The object for the eyepiece is formed by the objective; it is the image formed by the objective.   (frame 21)

12. to gather as much light as possible so that dim objects may be seen   (frame 23)

13. 50 (magnification $= f_{objective} \div f_{eyepiece} = 100$ cm $\div$ 2 cm)   (frame 22)

14. 300 (the product of the two magnifications)   (frame 21)

15. 22.22 cm. Solution: $\frac{1}{f} = (n - 1)(\frac{1}{R_1} + \frac{1}{R_2})$
$$= (1.6 - 1) \, (\tfrac{1}{20} + \tfrac{1}{40})$$
$$= 0.045 \qquad \text{(frame 4)}$$

16. 200 cm from the lens. Solution: $\frac{1}{f} = \frac{1}{p} + \frac{1}{q}$
$$\frac{1}{22.22} = \frac{1}{25} + \frac{1}{q}$$
$$\frac{1}{q} = 0.0050 \qquad \text{(frame 10)}$$

17. 33.9 cm. Solution: $\frac{1}{f} = (n - 1)(\frac{1}{R_1} + \frac{1}{R_2})$
$$= (1.52 - 1)(\tfrac{1}{15} + \tfrac{1}{-100})$$
$$= 0.0295 \qquad \text{(frame 5)}$$

18. 176 cm from the lens. Solution: $\frac{1}{f} = \frac{1}{p} + \frac{1}{q}$
$$\frac{1}{33.9} = \frac{1}{42} + \frac{1}{q}$$
$$\frac{1}{q} = 0.00569 \qquad \text{(frame 10)}$$

19. 11.11 cm from the lens.   Solution: $\frac{1}{f} = \frac{1}{p} + \frac{1}{q}$
$$\frac{1}{20} = \frac{1}{p} + \frac{1}{-25}$$
(The minus sign appears because the lens is being used as a magnifier, so the image is on the same side of the lens as the object.)
$$\frac{1}{p} = 0.09 \qquad \text{(frame 13)}$$

20. diverging, because the focal lengh is negative   (frame 15)

CHAPTER TWENTY-TWO

Light as a Wave

Prerequisite: None, but Chapters 10 and 12 suggested

In the beginning of our study of light (Chapter 18), we considered the nature of light—is it wave or particle? Chapter 19 focused on the particle nature of light, but the phenomenon of refraction (studied in Chapters 20 and 21) was explained with a wave model. In this chapter we will study additional phenomena that rely on a wave explanation.

A study of this chapter will enable you to:

- define diffraction;

- demonstrate how light bends around corners;

- indicate the effect of constructive and destructive interference of light waves;

- explain the cause of the pattern resulting from double slit interference;

- state the relationship between a diffraction grating and double slit interference;

- explain the colors of soap bubbles in terms of wave theory;

- indicate how crossed polarizers reduce the amount of light that passes through them;

- state the conditions necessary to observe diffraction of light;

- use the double slit interference equation to determine the location of bright spots resulting from double slit interference (optional);

- state the effect of wavelength on the double slit interference pattern (optional);

- use the diffraction grating equation to determine the location of bright spots in the diffraction pattern (optional).

DIFFRACTION

1.   In our everyday living we assume light travels in a straight line. We are aware of exceptions—when light goes through a lens or prism—but most of us assume that when light travels through still air, it goes straight. We align objects by sighting along them. We aim a

gun by sighting along it. These activities depend upon light not bending its path. But light does bend around corners! It only bends a little, however, and only if we know where to look will we notice it.

The best way to see the bending of light around corners (called "diffraction") is to use a razor blade to cut a short slit in a piece of paper, and then look through the slit at a distant source of light. Hold the slit close to your eye and open and close the slit by flex-

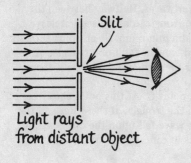

ing the paper. You will see the light apparently spread out perpendicular to the slit. The figure at the left illustrates the light passing through a slit perpendicular to the plane of this page and spreading out before entering the eye.

(a)   Is diffraction the result of light going from one material into another like refraction? _____

(b)   Is diffraction the result of the bouncing of light from a surface (like reflection)? _____

– – – – – – – – – – – – – – – – – –

(a) no; (b) no

2.   The phenomenon of diffraction is most easily explained by a wave model of light. If light consisted of particles, we would expect it to pass on through an opening with no spreading. Waves, however, do show spreading after passing through an opening. Sound waves spread after passing through a door or window, and water waves will do the same thing after being passed through small openings. If you try the latter by using two sticks to make a $\frac{1}{4}$ to $\frac{1}{2}$ inch opening for water waves in a pan, you will find that the smaller the opening the more the spreading. This corresponds to what we see in the case of light—the smaller the opening, the greater the diffraction effect.

In addition, sound waves and water waves bend around objects in their path. If light did this on a large scale, there would be no shadows, but if a small object is placed in a beam of light, such bending can be observed. (It is best observed with a laser—the bright light of the laser being easy to see after it curves around the object.)

(a)   What term refers to the bending of light around corners? _____

(b)   What amount of diffraction would result if light waves were particles instead?

_____

(c)   Is the bending more obvious when the opening is large or small? _____

– – – – – – – – – – – – – – – –

(a) diffraction; (b) none; (c) small

DOUBLE SLIT INTERFERENCE*

3.   In the figure in this frame waves are shown spreading from a source (L) and then striking two slits (M and N). After passing through the slits, the light again spreads out.

_____

*You may want to look over frames 3–5 in Chapter 12 before studying this. Those frames deal with a similar case for sound waves.

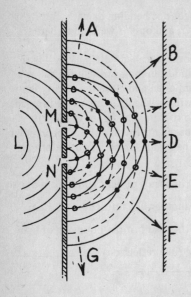

But notice that the light from one of the slits overlaps the light from the other. The line from B to F represents a screen upon which the light falls. At the center of the screen, at point D, the wave crests (represented by the semicircles) from one slit overlap the wave crests from the other slit. Visualize this as occurring with water waves. Halfway between the crests of a wave are troughs (dashed semicircles) where the wave is at a minimum. So at point D crests fall upon crests and troughs fall upon troughs. And the light is stronger there than it would be if there were only one slit.

Now look at point C. At this point the crests of one wave fall upon the troughs of the other and the troughs of the first wave fall upon the crests of the second. This results in cancellation of the light at those points.

At points where the two waves add so as to reinforce each other we say that *constructive* interference is taking place. At points such as C, where the waves cancel one another, we say that there is *destructive* interference.

Referring back to the figure, indicate whether there is constructive or destructive interference for each of the points.

A _____   B _____   C _____

D _____   E _____   F _____

G _____

— — — — — — — — — — — — — —

constructive: B, D, and F
destructive: A, C, E, and G

4.   To observe double slit interference in practice, the slits must be very small, like those you used to observe diffraction. In addition, the slits must be very close together, about 0.1 mm or less. Since a regular light bulb does not produce intense enough light to pass through such narrow slits and still illuminate a screen visibly, one normally observes "double slit interference" either by substituting the retina of the eye for the screen (by holding the slits just in front of the eye), or by using a laser as a source of light.

Suppose you manage to get a double slit interference pattern on a screen. What

pattern will you see? _____

— — — — — — — — — — — — — —

bright and dark spots across the screen

Optional

5.   Several variables influence the distance between consecutive bright spots on a screen illuminated by double slit interference. Perhaps the most obvious variable is the distance from the slits to the screen (distance d in Figure 22.1). Since the pattern

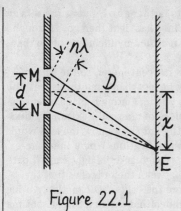

Figure 22.1

spreads from the slits, we would expect that the larger that distance D is, the farther apart will be the bright spots on the screen. The other variables involved are the wavelength of the light (λ) and the separation of the slits (d).

Figure 22.1 shows lines from each of the slits to a spot (E) on the screen. If the spot is to be bright, the distance from M to E must be greater than the distance from N to E by exactly one wavelength (or two wavelengths, or three, and so on). This extra distance is labeled in the figure as $n\lambda$ (n indicating 1, 2, 3, and so on). The small triangle with $n\lambda$ as one of its sides is similar to the large (dashed-line) triangle with x as one of its sides. Thus, making a slight approximation, we can write a ratio of length of sides:

$$\frac{n\lambda}{d} = \frac{x}{D}$$

Suppose we have two slits 0.1 mm apart and shine light through them on to a screen 5 meters distant. The first bright spot beyond the center one is measured to be 2.0 cm from the center spot.

(a)   What is the wavelength of the light used? (Hint: First put all variables in meters.)

_____

(b)   The second bright spot will be how far from the center spot? (This is found by

using $n = 2$.)_____

– – – – – – – – – – – – – – – – –

(a)   400 nanometers or $4 \cdot 10^{-7}$ meters. Solution:

$$\frac{n\lambda}{d} = \frac{x}{D}$$

$$\lambda = \frac{xd}{Dn}$$

$$= \frac{(.02)(.0001)}{(5)(1)} \quad \text{(all distances in meters)}$$

(b)   .04 meters or 4 cm. Solution: From the equation:

$$x = \frac{\lambda Dn}{d}$$

$$= \frac{(4 \cdot 10^{-7})(5)(2)}{.0001}$$

THE DIFFRACTION GRATING

6.   The double slit interference pattern is different for different wavelengths, and it could be used to analyze light (similar to the way a prism is used), but the pattern is normally fairly dim and the bright spots are very close together. Making the slits closer

together spreads out the pattern, putting the bright spots farther apart. But as the slits are made closer together, they must also be made smaller. Since less light then goes through them, the pattern will be dimmer. This situation can be rectified by including more than two slits.

In a *diffraction grating* there are thousands of slits per centimeter, so the slits are less than a thousandth of a centimeter apart. (They may be less than a ten-thousandth of a centimeter apart.) Such a grating is made by mechanically scratching a piece of glass with extremely fine lines. The figure at the left shows how the slits would appear under high magnification. Where there are scratches, the glass is rough and the light cannot pass through. Thus it is between the scratches that the light passes—these are the "slits." Diffraction gratings spread the light so much that the first bright spot may be at an angle of 30 degrees or more from center. They are used in place of prisms to analyze light, since each color has its bright spot at a different angle.

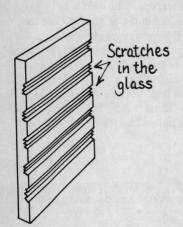

Scratches in the glass

The diffraction grating discussed above works by transmission of light. Diffraction can also be observed in reflected light, however. If you look at an angle across a record album, you can see color in the reflection. This color results from interference of the light which bounces from the various slits on the record, and is the result of essentially the same process as occurs with the standard diffraction grating.

(a)   What is it about a diffraction grating which causes its pattern to be spread out more than the double slit pattern?_____

(b)   How is it that the diffraction grating pattern is brighter than the double slit pattern?

_____

– – – – – – – – – – – – – – –

(a)   The slits are closer together.
(b)   The diffraction grating has many slits for the light to come through.

Optional

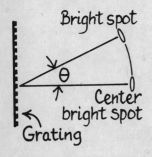

Bright spot

θ

Center bright spot

Grating

7.   The diffraction grating spreads the light at a far greater angle than simple double slits. For this reason we write the equation in terms of the angle of spread as follows:

$$n\lambda = d \cdot \sin \theta *$$

where d is again the distance between adjacent slits, and the angle (θ) is shown in the figure at the left.

_____

*"Sin" is the abbreviation for sine. The sine of any angle between 0° and 90° can be found in Appendix III.

Using the sine table in Appendix III, calculate the angle of the second bright spot produced by light of wavelength 400 nanometers (1 nanometer = 10^{-9} meters) passing through a grating with 8000 lines per centimeter. (Hint: You must first find the distance d between slits.) _____

– – – – – – – – – – – – – – – –

$40°$. Solution: If there are 8000 lines per centimeter, the distance between adjacent lines is $\frac{1}{8000}$ cm, or $1.25 \cdot 10^{-4}$ cm, or $1.25 \cdot 10^{-6}$ m. Putting all distances in meters:

$$\sin \theta = \frac{n\lambda}{d}$$

$$= \frac{(2)(400 \cdot 10^{-9})}{1.25 \cdot 10^{-6}}$$

$$= .64$$

THIN FILM INTERFERENCE

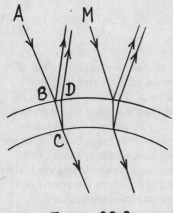

Figure 22.2

8.   In order to produce interference of light, a beam of light must somehow be broken into two beams and then recombined so that one of the separated beams travels farther than the other. We saw that this could be done by the use of double slits or a diffraction grating. Another example of interference is the production of the colors in soap bubbles. Figure 22.2 illustrates a greatly magnified section of a soap bubble.

A light ray is shown coming from point A and striking the bubble at B. When it hits the surface of the soap film, some of the light is reflected from it, but some continues through the surface. When this latter portion of the light hits the inner surface (at C), some of it is again reflected and some continues into the bubble. We will concern ourselves here with the reflected portion. It goes back toward the outer surface where some of it emerges from the bubble (at D) and recombines with the portion of the original beam which reflected from the surface upon first encountering it. These two waves overlap.

(a)   Which part of the light has gone the greater distance? _____

_____

(b)   What condition will create constructive interference? _____

_____

– – – – – – – – – – – – – – –

(a) the one that passed to the inner surface; (b) when the greater distance is a multiple of the wavelength.

9.   Now consider the beam leaving point M in Figure 22.2. It strikes the bubble where it is thicker. If this thickness is such that the extra path distance of the portion of the light entering the soap is $\frac{1}{2}$ (or $1\frac{1}{2}$ or $2\frac{1}{2}$ or $3\frac{1}{2}$, and so on) wavelength, the two waves will interfere destructively upon recombining. That particular wavelength of light will therefore not be seen to reflect at that point on the bubble.

(a)   What do different wavelengths of light correspond to? _____

(b)   If white light (containing all wavelengths) strikes the bubble, what color(s) would you expect to see reflected? _____

(c)   What causes the changing colors as you watch the bubble over a period of time?

_____

(d)   Consider an oil spot on the pavement of a parking lot. What causes the colors you see in it? _____

_ _ _ _ _ _ _ _ _ _ _ _ _ _ _ _

(a) different colors; (b) all colors, at different places on the soap bubble; (c) the walls of the bubble change thickness; (d) interference resulting from the thin film of oil

POLARIZATION

X

Y

10.   When water waves move across the surface of water, the primary motion of the water is up and down. By the very nature of the wave, they must vibrate vertically. If you vibrate the end of a stretched rope or spring, however, you can cause waves to vibrate either vertically or horizontally. Light waves are similar. Figure X shows how we might think of the vibration of a light wave as it is coming straight toward us. The arrows are used to represent vibration in all directions. Such a wave is termed *unpolarized*.

If unpolarized light is passed through a polarizer, all waves are absorbed except those vibrating in one particular plane, such as illustrated in Figure Y.

(a)   Now consider Figure 22.3 on the next page. Is the wave at A polarized or unpolarized?

_____

(b)   What does the polarizer do? _____

(c)   The polarizer at D is at a 180° angle to the one at B. What will be its effect on the light polarized by B? _____

_____

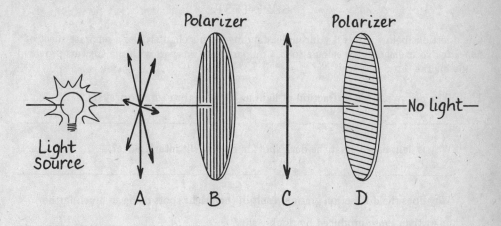

Figure 22.3

(a) unpolarized; (b) it allows only waves vibrating in the vertical plant to get through;
(c) doesn't allow vibrations in that plane through at all, so no light gets through

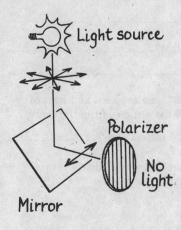

11.   Figure 22.3 showed the use of two polarizers to eliminate light. The figure at left shows another method. Here light reflects from a shiny surface, which polarizes light (at least partially) in the horizontal plane. In the figure, the reflected light is shown going toward a polarizer which allows vertical vibrations to pass. Thus, if the reflection results in perfect polarization, none of the reflected light gets through. This is the secret of polarizing sunglasses. The glare which reuslts from reflection from smooth surfaces such as car bumpers and water surfaces is at least partly eliminated by such glasses. To see this effect, rotate polarized glasses in front of your eyes as you look through them at bright reflections from various objects.

How is it that polarized sunglasses don't prevent you from seeing at all? _____

_____

not all light is polarized

SELF-TEST

The questions below will test your understanding of this chapter. Use a separate sheet of paper for your diagrams or calculations. Compare your answers with the answers provided following the test.

1.   Why do we not see diffraction of light as it passes through doorways? _____
_____

2.   Why is light canceled at the dark spots in double slit interference? _____
_____

3.   Why does the diffraction grating result in the bright spots being so much farther
apart than those produced by double slits? _____
_____

4.   Refer to Figure 22.2. How much farther must the beam entering the soap film travel
if a bright area is to be seen? _____
_____

5.   Why are different colors seen at different places on a soap bubble? _____
_____

6.   How does polarized light differ from unpolarized light? _____
_____

7.   *Optional.* Suppose double slits are spaced $1.33 \cdot 10^{-4}$ meters apart and light of
500 nanometers wavelength shines through them. What distance from the center
spot will the *second* bright spot appear on a screen 4 meters away? _____

8.   *Optional.* A grating with 12,000 lines per centimeter is used with light of wavelength 650 nanometers. What is the angle of the first bright spot?

_____

9.   *Optional.* What effect on the double slit pattern will result from an increase in the
wavelength of light? _____
_____

Answers to Self-Test

If your answers do not agree with those given below, review the frames indicated in parentheses before you go on to the next chapter.

1.   Diffraction of light only occurs to any degree when the opening through which the light goes is very small.   (frames 1-2)

2.   The troughs of one of the beams meet the crests of the other and vice versa. Crest cancels trough, and a dark spot results.   (frame 3)

3.   The slits of a diffraction grating are much closer together than it is practical to make double slits.   (frame 6)

4.   It must travel a whole number of wavelengths farther.   (frame 8)

5.   At any given place, only one wavelength will show perfect constructive interference. All other wavelengths will interfere destructively to some degree.   (frames 8-9)

6.   In polarized light, the wave vibrates in only one plane, while unpolarized light has waves vibrating in all directions perpendicular to the direction of travel. (frame 10)

7.   0.03 meter, or 3 cm.   Solution:

$$\frac{n\lambda}{d} = \frac{x}{D}$$

$$x = \frac{n\lambda D}{d}$$

$$= \frac{(2)(500 \cdot 10^{-9})(4)}{1.33 \cdot 10^{-4}} \quad \text{(in meters)} \quad \text{(frame 5)}$$

8.   51°.   Solution: $\frac{1}{12,000} = 8.33 \cdot 10^{-5}$, which is the grating space in centimeters. This is $8.33 \cdot 10^{-7}$ meters.

$$\sin \theta = \frac{n\lambda}{d}$$

$$= \frac{(1)(650 \cdot 10^{-9})}{8.33 \cdot 10^{-7}}$$

$$= 0.780$$

The table of Appendix III yields the answer.   (frame 7)

9.   The spots will get farther apart. This follows from the equation:

$$\frac{x}{D} = \frac{n\lambda}{d}$$

As the wavelength (λ) gets larger, so does distance x, the distance between slits. (frame 5)

CHAPTER TWENTY-THREE

Color

No Prerequisites

Visible light is electromagnetic radiation with wavelengths between 400 nanometers and 700 nanometers. (A nanometer is 10^{-9} meters.*) Within this range, a difference in wavelength appears to us as a difference in color.

When you complete your study of this colorful chapter, you will be able to:

- state the relationship of color to wavelength;

- state which colors have the longest and which have the shortest wavelength;

- name the additive primary colors;

- specify the result when given colors are combined additively;

- specify the result when given colors are combined subtractively;

- name the subtractive primary colors;

- specify the result when additive and subtractive processes are combined;

- indicate the reason that paints don't combine in real life accordin to the theory in this chapter;

- specify the method of color combination used on a color television screen.

THE VISIBLE SPECTRUM

1.   Figure 23.1 shows the order of the colors in the visible spectrum, from violet light of shortest wavelength to red at the longest wavelength.

It is important to realize that the range of each color indicated is arbitrary. Red fades gradually into orange, and different people might put the dividing line between the two colors in different places. In fact, the number of colors which you might list across the spectrum is arbitrary. You might want to list aqua as a separate color between blue and green, or indigo between blue and violet.

---

*See Appendix I for a discussion of powers of 10.

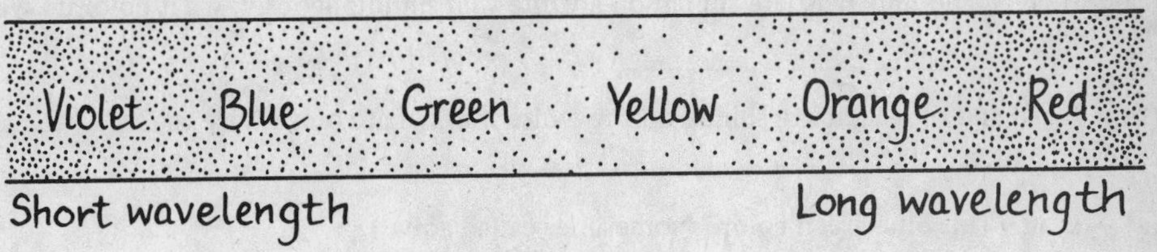

Violet   Blue   Green   Yellow   Orange   Red

Short wavelength   Long wavelength

Figure 23.1

Which has the shorter wavelength, a blue-green or a yellow-green color? _____

_____

blue-green

2.   White light contains all wavelengths of the spectrum, so it contains all colors. Yet, when white light shines on a banana, the light entering your eye from the banana is yellow. This is because the surface of the banana absorbs some colors from the white light and reflects primarily yellow.

What causes an unripe banana to appear green?

Blue	Green	Red

Short   Long
wavelength   wavelength

_____

The surface absorbs other colors, reflecting green.

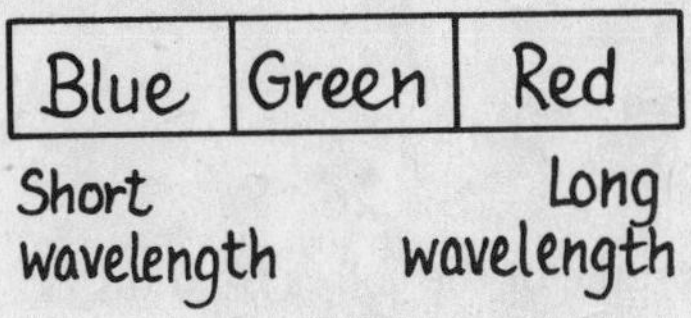

Figure 23.2

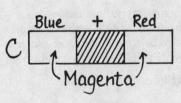

A [////] Green + Red Yellow

B Blue + Green Cyan [////]

C Blue   +   Red [////]
   Magenta

Figure 23.3

3.   Although the color green is shown as a small portion of the visible spectrum, if light of the entire central third of the spectrum strikes the eye, we see the color green. Thus, for convenience, we may divide the spectrum into thirds and name them as shown in Figure 23.2. The long-wavelength third appears red to us, although it actually includes wavelengths which by themselves appear orange. The short wavelength third appears blue.

Now suppose that we have a wall which normally appears white (because it reflects all colors of light). If we shine light which contains only the central portion of the spectrum on this wall, the wall will appear green. What if we shine green light on this wall at the same time that we are shining red light on it? In this case, two-thirds of the spectrum is being reflected into our eye from the wall. (See Figure 23.3, part A.) This two-thirds of the spectrum appears yellow to us. (You will note that on Figure 23.1, which shows all colors of the spectrum, yellow appears in the center of this range.) Refer to the remainder of Figure 23.3.

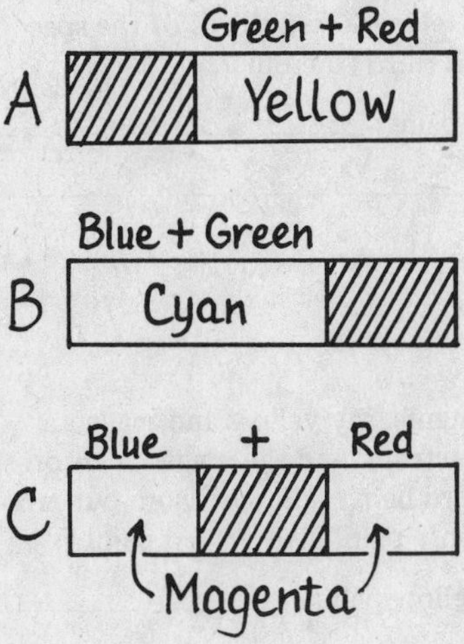

(a)   If green and blue light are shined on a white wall simultaneously, what color do we see? _____

(b)   If red and blue light are shined on the white wall, what color do we see? _____

— — — — — — — — — — — — — — —

(a)   cyan (This blue-green color is sometimes called aqua.)
(b)   magenta (This is a purplish red.)

4.   In the last frame, we formed new colors by actually adding colors together. This process is called an *additive* process. On the screen of a color television set are small dots (or rectangles) of the three additive primary colors, small enough that we cannot see them individually from a normal viewing distance. The colors we see on the screen are produced by selectively lighting up these three primary colors. Light from one third of the spectrum is combined with light from another third to produce other colors.

(a)   What are the three additive primary colors? _____

(b)   How would yellow be produced by the additive process on a color TV set?

_____

— — — — — — — — — — — — — — —

(a) red, green, and blue; (b) lighting up the green and the red dots

5.   An object appears a certain color because it absorbs some of the colors from the spectrum and reflects others. It *subtracts* out some of the colors of the spectrum and we see only what is left. In frames 3 and 4 we discussed forming new colors by *adding* lights together. Pigment colors combine differently, by a subtractive process. The primary colors which will allow us to obtain all other colors by mixing pigments are yellow, cyan, and magenta, as indicated in Figure 23.3. Each subtractive primary contains two thirds of the spectrum. Yellow paint, for example, reflects the long wavelength two-thirds of the spectrum and absorbs—or subtracts out—the short wavelength third (the blue).

(a)   What part(s) of the spectrum are absorbed by cyan paint? _____

(b)   What part(s) are reflected? _____

— — — — — — — — — — — — — —

(a) the red third; (b) the blue and green thirds

6.   Now consider what happens when we mix two pigments, say yellow and magenta paint. The yellow pigment absorbs the blue end of the spectrum, and the magenta absorbs the central portion, the green. There is therefore still a third being reflected from our mixture of paint, the red. Thus a mixture of yellow and magenta paint gives us red paint.

What color is produced by a mixture of cyan and yellow paint? _____

— — — — — — — — — — — — — —

green (The yellow absorbs the blue third, reflecting the other two. The cyan absorbs the red third, reflecting the other two thirds. Thus between the two pigments, both the blue and red ends of the spectrum are absorbed. This leaves only green, the central portion, to be reflected.)

UNDER COLORED LIGHTS

7.   You have seen that an additive process results from colored lights, and a subtractive process results when pigments are combined. These two processes work together when colored objects are viewed under colored lights. For example, suppose a yellow object is placed in red light. The object will appear red because a yellow object is capable of reflecting red light. (See Figure 23.3, part A.) This same object under blue light will appear black, however, because it cannot reflect blue light. Since no light is reflected, it appears black to us.

(a)   Now consider the yellow object under cyan light. What color does it appear?

_____

(b)   What color does a red object appear under magenta light? _____

(c)   What color does a cyan object appear under magenta light? _____

– – – – – – – – – – – – – – – –

(a)   green (The object can reflect the green and red thirds of the spectrum. The cyan light shines the blue and green thirds onto it, but the blue is not reflected. The green is therefore left.)
(b)   red (The magenta light includes the red third.)
(c)   blue (Blue is the only portion in the magenta light which can be reflected by the cyan object.)

WHY ISN'T IT THIS PERFECT IN PRACTICE?

8.   In order to do the analysis we did in this chapter, we made some simplifying assumptions. If you go out and buy red paint and analyze its reflected light, you will find that its spectrum does not fit perfectly into the short-wavelength third of the visible spectrum. It may even reflect some light from the violet end. In general, actual pigments reflect a complicated spectrum, and one must know this spectrum well in order to predict accurately what color will appear when two pigments are mixed. If we understand the simple processes, however, we can better understand what happens in the more complicated real world.

(a)   By what process do colors in lights combine? _____

(b)   By what process do colors in pigments combine? _____

(c)   By what process do colors in TV screens combine? _____

– – – – – – – – – – – – – – – –

(a) additive; (b) subtractive; (c) additive

SELF-TEST

The questions below will test your understanding of this chapter. Use a separate sheet of paper for your diagrams or calculations. Compare your answers with the answers provided following the test.

1.   What is different concerning light waves which result in the phenomenon of color?
      _____

2.   Name the additive primaries. _____

3.   What color results when red and blue lights are shined on a white wall? _____

4.   What color results form the addition of red and green light? _____

5.   Name the subtractive primaries. _____

6.   What color paint results from the mixing of magenta and cyan paints? _____

7.   Are colors produced on a television screen by an additive or a subtractive process?
      _____

8.   What color will a red object appear under yellow light? _____
      Under green light? _____ Under magenta light? _____

9.   What color will a magenta object appear under green light? _____
      Under blue light? _____ Under yellow light? _____

10.   Why do the examples of the additive and subtractive processes used in this chapter
      sometimes not work well in practice? _____
      _____

11.   If cyan and red light are shined on a white wall, what color will the wall appear?
      _____ Magenta and green on the white wall? _____

Answers to Self-Test

If your answers do not agree with those given below, review the frames indicated in parentheses before you go on to the next chapter.

1.   wavelength (or frequency)   (frame 1)

2.   red, green, and blue   (frames 3-4)

3.   magenta   (frames 3-4)

4.   yellow   (frames 3-4)

5.   yellow, cyan, and magenta   (frame 5)

6. blue (because it is common to both pigments)   (frame 6)

7. additive   (frame 4)

8. red; black; red   (frame 7)

9. black; blue; red   (frame 7)

10. Actual pigments cannot be made to reflect exactly one-third or two-thirds of the spectrum; they are more complicated than this; so the results are more complicated. (frame 8)

11. white; white   (frames 2–3)

PART SIX

The Physics of Relativity and the Nucleus

The last part of this book considers two of the more recently established branches of physics—relativity and nuclear physics. We spend only one chapter on the first of these, but that will suffice for us to delve into some of the strange predictions which result from this world-view first presented by Albert Einstein in 1905.

Nuclear physics has somehow gained a reputation as being an extremely difficult subject, but we will see that its basic principles are no more difficult to master than those of any other branch of physics. After a study of radioactivity, we will examine a few of the many practical uses of radiation which have become widespread in today's world. Your study of physics ends with nuclear energy, that energy which is seen dramatically—and horribly—in the atomic bomb, that energy which today supplies a small fraction of the world's energy needs (from fission power plants), and that energy which may some-day (from fusion power plants) provide a clean, safe, and almost unlimited energy supply.

CHAPTER TWENTY-FOUR

Special Relativity

No prerequisites, but includes references to Chapter 18, frames 13–14

Most people have heard of Albert Einstein and his theory of relativity, but few realize that Einstein formulated two theories of relativity: the special theory and the general theory. We will limit ourselves to the special theory of relativity in this Self-Teaching Guide. First, however, we will consider a theory of relativity which predated Einstein by more than two centuries.

Completion of this chapter will enable you to:

- differentiate between Newtonian and Einsteinian relativity;

- specify the frame of reference for speed in a given situation;

- combine velocities in Newtonian relativity and in special relativity (given the formula);

- relate the Michelson-Morley experiment to the ether;

- explain the effect of speed on time, mass, and length of the object in motion;

- explain the meaning of $E = mc^2$;

- state the frame of reference for the speed of sound;

- state the frame of reference for the speed of light;

- state the reason for postulating the existence of the ether.

NEWTONIAN RELATIVITY

1.   When Isaac Newton stated his laws (see Chapter 1), he included an hypothesis that has come to be known as Newtonian relativity. According to Newton, all laws of mechanics that are true in one system are also true in any system that is moving at a constant velocity relative to the first. This principle means that you cannot determine by any mechanical experiment (done entirely within the system) whether that system is moving or not. In fact, the speed of one system can only be stated with reference to another

system. For example, when we are traveling along a road at 50 miles per hour, this means that we are moving 50 mph *relative* to the earth. And a sign beside the road is moving 50 mph *relative* to us. Because we are so used to relating everything to the earth, we seldom think of things moving relative to us. Young children riding in a car, however, say that trees and signs are moving—they tend to refer all motion to themselves.

From the standpoint of the universe, we might relate motion to the sun, and it is clear that the sign by the road (and the whole earth) is moving relative to the sun. The point is that anytime we say that something is moving (or standing still), we are relating its motion to some other object.

(a)   When two boats are far out on the ocean, what is most apparent, the boats' motions

relative to the sun or to one another? _____

(b)   Suppose you are riding in a plane and a child in the seat behind you hits you in the head. What is more important to you, his hand's motion relative to you or to the

earth? _____

––––––––––––––––––––

(a) to one another; (b) to you

2.   When you are going down the highway at 55 miles per hour, a car may pass you from behind, going by at 10 miles per hour relative to you. Its speed relative to the earth would be 65 miles per hour. If a passenger in your moving car throws a ball from the back seat to the front, and the speed of the ball (relative to the car) is 20 mph, its speed relative to the earth is 75 mph. But if you throw the ball from the front to the back seat at this speed, its speed relative to the earth would then be 35 mph. The point of all this is that according to Newtonian relativity, speeds can be added in a "normal" way. We will see that according to Einsteinian relativity, we must add speeds differently in certain circumstances.

Suppose you are in your car, traveling at 50 mph. What are the speeds of the vehicles below, relative to *you?*

(a)   A car following you at 50 mph. _____

(b)   A car traveling in the opposite direction at 50 mph. _____

(c)   A car you pass that is going 30 mph. _____

––––––––––––––––––––

(a) 0; (b) 100 mph; (c) 20 mph in the opposite direction you are going

THE SPEED OF SOUND

3.   In Chapter 11 it was stated that the speed of sound is about 1100 feet per second, relative to the air. One cannot say that this speed is relative to the earth because the air may well be moving across the surface of the earth, and sound travels *through air* at a speed of 1100 ft/s.

Suppose you shout when a strong wind is blowing. Who will hear you first, someone

standing upwind or someone standing the same distance downwind?_____

––––––––––––––––––––

the person standing downwind

LIGHT AND THE ETHER

4.   The speed of light is 186,000 miles per second. Since light will travel in a vacuum, air cannot be the frame of reference for the speed. To solve the problem of the correct frame of reference for the speed of light, physicists of the late 1600s assumed that a substance they called the ether (or luminiferous ether) permeated all space, and was the frame of reference they sought. The ether was not a material in the sense of other materials, for it existed in a vacuum, and the earth and planets moved right through it without resistance. It was thought to be the absolute frame of reference to which all motion in the universe could be related.

Is there an ether? The best answer to this question is to ask what light travels in if there is no ether. But because of ether's unusual properties, it would be very difficult to detect. In 1887, Albert A. Michelson and Edward W. Morley set out to detect this ether and to determine the earth's motion relative to it.

If you haven't already done so, you might find it useful to refer to frames 13 and 14 in Chapter 18. In that section, we see that Michelson and Morley were unable to detect the ether and had to conclude that there is no reason to think that it exists. But we are then left with a dilemma concerning the frame of reference in which light travels at its 186,000 miles per second. Every motion must be referenced to something, but we don't know the frame of reference for the speed of light.

(a)   Does Newtonian relativity give us a frame of reference for light? _____

(b)   What did the Michelson-Morley experiment tell us about the frame of reference for

light? _____

— — — — — — — — — — — — — — —

(a) no; (b) it wasn't the ether

EINSTEIN'S POSTULATES

5.   In 1905, Einstein advanced his theory of relativity, which is based on two postulates. The first postulate is:

> If two systems are moving relative to one another at a constant velocity, all laws of nature are the same in the two systems.

This means that if you are moving along in a car on a long straight road at a constant speed, you could do nothing entirely inside the car that would show that you were indeed moving. If you throw a ball upward, it will come down in exactly the same way as it would if the car were sitting still at a traffic light. Newtonian relativity predicted similar results in mechanical experiments such as these. It was the fields of light and electricity where Einstein expanded the idea of relativity. His first postulate meant that there would be no experiment whatsoever that could tell you whether or not you were in motion— there is no such thing as absolute motion, but only motion *relative* to something else.

Consider, then, two spaceships moving toward one another. (We use spaceships in the examples here so that we can better imagine the great speeds which we must discuss.) A person in each ship has a flashlight and a device that will allow the measurement of the speed of light from the flashlight. If both experimenters make this measurement, both will get exactly the same speed for the light, whether or not the flashlight is pointing in the direction the ship is going. Again, it is impossible to determine speed (or motion) from any experiment in the ship.

(a)   When a person is in a rapidly moving spaceship, what system is all motion detected

relative to? _____

(b)   What does Einstein's first postulate say about the laws of nature in the spaceship

compared to those on earth? _____

— — — — — — — — — — — — — — — —

(a) the spaceship; (b) the same, if at a constant velocity

6.   Now suppose a person in one ship shines a light toward the other ship and the person in that ship measures the speed of the light coming toward it. Assume that the ships are moving together at the fantastic rate of 1,000 miles per second. The person in the first ship saw the light going away at 186,000 miles per second. The other ship is moving toward the first at 1000 mps, so someone in the second ship might be expected to measure the speed of the oncoming light to be 187,000 mps (186,000 plus the relative speed of the ships). Einstein's second postulate refutes this, however:

> The speed of light (in a vacuum) is independent of the motion of the observer.

According to this postulate, the person in the second ship also measures the speed of light as 186,000 mps! In fact, if somehow you could travel at the speed of 186,000 mps and measure the speed of a beam of light coming from the other direction, you would get the same 186,000 mps, rather than twice that great, as Newtonian relativity predicts.

Still, Einstein's "laws" are only postulates. They really do not explain what happens, but simply state a rule that corresponds to the results of experiments. The true test of the usefulness of the theory of special relativity lies in following it through to its logical conclusions and testing its implications. There are a number of quite unusual effects implied, as we will see in later frames.

(a)   Would Newtonian relativity predict that the earth's rotation has an effect on the

speed of light? _____

(b)   Does Einstein's special relativity predict that the earth's rotation has an effect on

the speed of light? _____

(c)   During the Michelson-Morley experiment, the speed of light measured the same across their apparatus no matter in which direction the instrument was oriented.

Which relativity theory explains Michelson's and Morley's experiment? _____

(d)   Which of Einstein's postulates deals directly with the speed of light? _____

— — — — — — — — — — — — — — — —

(a) yes; (b) no; (c) Einstein's; (d) second

7.   According to Newtonian relativity, if you are traveling in a plane at a speed of 300 miles per hour relative to the earth, and you throw a ball toward the front of the plane with a speed of 60 miles per hour (relative to the plane), the ball's speed relative to the earth is 360 mph. This kind of result can be verified experimentally with high precision, and the theory of relativity had better not predict anything different. And it doesn't—

until a very great speed is reached. Say, for example, that a spaceship is traveling at 140,000 miles per second relative to the earth. (This is 0.75 c, or 75% of the speed of light, a ridiculously large speed even for a spaceship.) It passes another spaceship going the other way, and this ship is also going 140,000 mps relative to the earth. According to the Newtonian method of adding velocities, the ships are traveling 280,000 mps relative to each other. But the theory of special relativity says that speeds must be added according to the following formula:

$$u = \frac{v_1 + v_2}{1 + (v_1 v_2 / c^2)}$$

where u is the combined velocity, v_1 and v_2 are the velocities being added, and c is the speed of light. The relative speed of the spaceships is 179,000 mps according to this equation.

If you try a variety of speeds in this equation, you will see that you can never have a speed greater than the speed of light. The equation also holds for low speeds, but if you do a calculation using speeds such as in the airplane example above, you will find that the difference between Einstein's result and the "normal" result is insignificant. Newtonian methods suffice for everyday events, and the difference predicted by relativity is not normally measureable. It is not until speeds reach a significant fraction of the speed of light that special relativity effects become important.

Assume that you have two light beams going in opposite directions. Use the equation above to determine their combined speed relative to one another. _____

_ _ _ _ _ _ _ _ _ _ _ _ _ _ _

186,000 mps.   Solution: Substituting into the equation, but writing c for v_1 and v_2 instead of the longer 186,000:

Light beam

Flashlight

A (as seen by
   woman in ship)

B (as seen by
   outside observer)

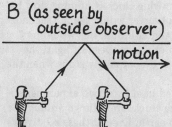

motion

Position when   Position
light left   when light
flashlight   returned

$$u = \frac{c + c}{1 + (c \cdot c/c^2)}$$

$$u = \frac{2c}{1 + 1} = \frac{2c}{2} = c$$

TIME DILATION

8.   Perhaps the strangest prediction of special relativity is that time passes at different rates under different circumstances. Refer to figure A at the left. We have a woman in a rocketship shining a flashlight beam straight up to a mirror on the ceiling and measuring the time for the beam to return to the flashlight. The time would be very short, but don't be concerned about how—or whether—it can be measured. Now suppose that this rocket is moving at high speed past an observer who can see the light beam leave the flashlight, reflect from the ceiling and return to the flashlight. But if the rocketship is moving by someone at a very great speed, this person sees the beam taking the path of figure B. It goes forward while going up and down.

Now the second postulate of special relativity states that the speed of light is the same when measured by each person. The man being passed by the rocket ship sees the light beam going a greater distance than does the woman in the ship. Figure A shows the view of the beam seen by the woman in the ship, while figure B shows how the beam looks to the observer on the outside. Note that the latter distance is greater, and recall that speed is distance divided by time.

(a)   If the speed is the same and the distance is different, what can you say about the

elapsed time for the two people? _____

(b)   If the woman in the ship measures the time to be one milli-microsecond, the observer outside might measure it to be 1.5 milli-microsecond. To the person outside the ship, which clocks move more slowly, his own or the ones in the ship?

_____

_ _ _ _ _ _ _ _ _ _ _ _ _ _ _

(a) must be different; (b) in the ship (since he saw more time pass on his clocks)

9.   The theory of relativity predicts that if one system is moving at a great speed with respect to another, time is seen to move slower in the system with great speed. This refers to time as measured by the motion of a light beam, by a clock, or by the biological processes of a living being. But now recall Einstein's first postulate; it says that it is impossible to say which of the systems is truly moving, but only that each is moving with respect to the other. This means that to a person in each system, a clock in the other system is moving at a slower rate. And, strange as it may seem, this is just what is predicted by the theory of relativity.

Why don't we observe this slowdown of time in everyday life? Because the effect is far too small to be noticeable at ordinary speeds. Only when speed approaches the speed of light does the effect become significant. The equation which predicts how much time (t) passes in the moving system when a "normal" time (t_0) passes in the observer's system is:

$$t = \frac{t_0}{\sqrt{1 - v^2/c^2}}$$

In this equation v represents the speed of a moving system relative to the observer's system, and c is again the speed of light. If the spaceship is moving at half the speed of light, the equation tells us that the "stationary" observer would see 1.15 seconds pass on the spaceship for every 1 second on his own clock. The fast-moving clock is running slow by 15%. At $0.9c$, nine-tenths the speed of light, the outside observer would see 2.3 seconds pass in the spaceship for every second on his own clock. And at $0.99c$, this would be 7 seconds! As the speed of the spaceship gets closer and closer to the speed of light, the clock in the spaceship is seen to go slower and slower, until it completely stops when the spaceship reaches the speed of light.

These weird predictions of relativity have been tested in at least two experiments, both of which have verified the predictions. Very accurate clocks were flown around the earth in high speed planes and compared to clocks left on earth. Even at these relatively slow speeds, the difference was detected. Radioactive cosmic rays coming at a high speed toward earth are seen to have a different half-life* than the same particles at rest on earth, the difference being just as predicted by the theory of relativity.

*See Chapter 25, frames 8-9 for the definition of half-life if you are interested.

Consider the spaceship and observer again. If the woman on the spaceship were measuring time, which clocks would she see as slower? _____

– – – – – – – – – – – – – – – –

those on earth

LENGTH CONTRACTION

10.   Another prediction made by the theory of relativity is that as an object reaches a high speed with respect to the observer, a measurement (by the observer) of the length of the object would show it to be shorter than it was when at rest. The equation giving the apparent length of a moving object is:

$$\ell = \ell_0 \sqrt{1 - v^2/c^2}$$

Here ℓ_0 is the length of the object when it is at rest with respect to the observer. For example, consider a meter stick moving by you like a spear. If it moves at half the speed of light, it appears to be 0.87 meters long. At 90% of the speed of light, it appears 0.44 meters long, and at $.99c$, it appears as 0.14 meters long.
   Consider a person on a rocket ship carrying the meter stick.

(a)   What is the speed of the stick with respect to this person? _____

(b)   What length contraction will he see when the ship goes 90% of the speed of light?

   Why? _____

   _____

– – – – – – – – – – – – – – – –

(a) zero; (b) none, because the length contraction occurs only because of a relative motion of the observer and the object being measured

MASS INCREASE

11.   In Chapter 1, we defined mass as a measure of the amount of inertia an object has. We said that an object with more mass requires more force to accelerate it a given amount. Though we often associate mass with weight, they are distinct concepts, since if we take an object far enough from the earth, the object has essentially no weight, but it still has the same mass. The force of gravity diminishes as one moves away from earth, but the force required to accelerate the object stays the same.
   The theory of relativity tells us that as an object increases its speed, its mass increases. Thus the amount of force required to accelerate an object is greater as it nears the speed of light.
   Will it require more force to accelerate an object from 175,000 mps to 180,00 or

from 180,000 to 185,000 mps in the same time? Why? _____

_____

– – – – – – – – – – – – – – – –

180,000 to 185,000, because of the increase of mass due to the object's greater speed

12.   The equation for mass increase is similar to that of time dilation:

$$m = \frac{m_0}{\sqrt{1 - v^2/c^2}}$$

where m is the mass at speed v, and m_0 is the mass when the object is at rest with respect to the observer. At a speed 99% of the speed of light, the object has a mass 7 times greater than its "rest mass."

The increase of mass with speed is a common observation in nuclear physics, where nuclear particles commonly attain speeds near the speed of light. One method of nuclear physics is to deflect such particles with electric and magnetic fields, and such deflection allows physicists to measure the masses of the particles. The measured increase of mass with speed verifies the theory of relativity.

Suppose a hypothetical (and *very* small) passenger is riding on a nuclear particle that is moving past her buddy at near the speed of light.

(a)   Will the passenger see her clocks or the clocks of her buddy moving slower?

_____

(b)   From her point of view, who will appear thinner, her buddy or herself? _____

(c)   To her buddy, who appears thinner? _____

(d)   If the two could measure each other's mass, whose mass would appear less to the nuclear passenger? _____

(e)   Of the three predictions of special relativity—time dilation, length contraction, and mass increase—for which have we given an example of actual experimental verification? _____

– – – – – – – – – – – – – – – – –

(a) her buddy; (b) her buddy, assuming that they are both about equally thin when at rest with respect to one another; (c) she does; (d) her own mass, since he is moving at a great speed with respect to her; (e) time dilation and mass increase

MASS-ENERGY EQUIVALENCE

13.   The remaining major prediction of special relativity provides the most spectacular verification of the theory. One of the basic laws in science before the theory of relativity was that mass is conserved, that is, that mass can neither be created nor destroyed. In a chemical reaction, the mass of the resultant products is always the same as the mass of the original chemicals. We just saw, however, that mass does increase. But only when we expend the required energy to speed up an object. Thus we see also that the conservation of energy principle—the main topic of Chapter 2—must be altered. Energy can be destroyed, by changing it into mass.

Einstein's equation for this transfer of energy (E) into mass (m) is:

$$E = mc^2$$

where c is again the speed of light. This equation not only applies to a change of energy into mass, but to a change of mass into energy. Einstein's prediction was verified by Enrico Fermi in 1942 in the first nuclear reactor, and more spectacularly by the explosion of the first atomic bomb a few years later. The source of energy in both the reactor and the bomb is mass—changed into energy according to Einstein's formula. We will study this energy production further under the topic of nuclear physics in the next three chapters.

(a)   What must you do to the speed of an object to increase its mass? _____

_____

(b)   Would you speed up or slow down an object to convert mass into energy?

_____

— — — — — — — — — — — — — — — —

(a) increase its speed; (b) slow it down (This is not what causes the energy increase in a nuclear reaction, however.)

SELF-TEST

The questions below will test your understanding of this chapter. Use a separate sheet of paper for your diagrams or calculations. Compare your answers with the answers provided following the test.

1.   What is the principle of Newtonian relativity? _____

_____

2.   According to Newtonian relativity, if you are in a train going 60 mph and you throw a ball forward at 50 mph, what is the speed of the ball relative to the earth?

_____

3.   What is the frame of reference for the speed of sound? _____

4.   Why did we once believe that the ether existed? _____

_____

5.   One of Einstein's postulates deals with a comparison between two systems and is similar to Newtonian relativity. What does it add to Newtonian relativity?

_____

6.   Light speed is 186,000 mps (or $3 \cdot 10^8$ m/s). What is the frame of reference for this speed? _____

7.   Must ordinary velocities be added according to the theory of special relativity? Why or why not? _____

8.   Suppose a spaceship traveling at $0.5c$ (with respect to the earth) shoots a rocket in the forward direction at a speed of $0.7c$ (with respect to the spaceship). Indicate which is the rocket's speed with reference to the earth.

_____ (a)   less than $0.5c$

_____ (b)   between $0.5c$ and $0.7c$

_____ (c)   between $0.7c$ and c

_____ (d)   between c and $1.2c$

_____ (e) c

9.   Would the clocks on a spaceship zooming by us appear to us to be running slow or fast?_____

10.   Would the pilot of the fast-moving ship of the last question see our clocks running slow or fast? _____

11.   Would a spaceship moving by us at a fast speed appear longer or shorter than when it is standing still? _____

12.   What is the formula $E = mc^2$ concerned with? _____

Answers to Self-Test

If your answers do not agree with those given below, review the frames indicated in parentheses before you go on to the next chapter.

1.   Any law of mechanics which holds in one system also holds in any system which is moving at constant velocity with respect to the first.   (frame 1)

2.   110 mph   (frame 2)

3.   the air in which the sound is moving   (frame 3)

4.   It was thought necessary for light to travel in.   (frame 4)

5.   Newtonian relativity only referred to experiments in mechanics. Einsteinian relativity includes all possible experiments in all areas of science.   (frame 5)

6.   the measurer of the speed   (frame 6)

7.   No. The difference is only significant at speeds near the speed of light.   (frame 7)

8.   Answer c. (The velocity cannot be as great as the speed of light, but it is greater than the larger of the two speeds being added.)   (frame 7)

9.   slow   (frames 8–9)

10.   slow   (frames 8–9)

11.   shorter   (frame 10)

12.   change of mass (m) to energy (E) or vice versa   (frame 13)

CHAPTER TWENTY-FIVE
Radioactivity

Prerequisite: Chapter 5

The center of the atom is the nucleus, which is made up of protons and neutrons. The nucleus of an atom makes up about 99.97 percent of the mass of the atom. The radiations that come from some nuclei will be the major topic of this chapter. But before considering nuclear radiation, we must examine some differences which exist between nuclei.

When you complete your study of this chapter, you will be able to:

- name the component of the nucleus that differentiates one isotope of an element from another;

- name the component of the atom which determines which element the atom is;

- state the meaning of the atomic number of an element;

- state the meaning of the mass number of an isotope;

- interpret subscripts and superscripts in nuclear symbolism;

- state the make-up and charge of alpha particles, beta particles, and gamma rays;

- use the alpha emission formula to determine the resulting isotope in alpha decay;

- use the beta emission formula to determine the resulting isotope in beta decay;

- state how a nucleus changes when a gamma ray is emitted;

- given the half-life of a substance, state the amount remaining after any number of half-lives;

- explain the part played by radioactive series in the changing of unstable isotopes into stable isotopes;

- state the reason why radioactive substances become less radioactive as time passes.

ISOTOPES

1.   One element was originally distinguished from another because each had definite, distinguishing properties. We now know that these properties are due to the atom's electron structure, which is related to the number of electrons in the neutral atom. The number of electrons circling the nucleus of a neutral atom is equal to the number of protons in its nucleus. An atom with one proton in its nucleus is hydrogen, and an atom with two protons is helium, three protons—lithium. The number of protons is called the atomic number; it is the number shown above the element's symbol on the periodic table on page 62.

.   In 1912 it was discovered that atoms of the same element sometimes differ in mass. For example, some neon atoms have a relative mass of 20 and some a mass of 22. These different types of atoms of the same element are called *isotopes* of the element. After the neutron was discovered (in 1932) scientists found that the number of neutrons in the nucleus can vary, resulting in various isotopes of the same element. The mass number of an atom or an isotope is equal to the sum of protons and neutrons in the nucleus of that isotope.

Neon's atomic number is 10; it always has 10 protons in its nucleus.

(a)   The isotope neon-20 (mass number 20) has how many neutrons?_____

(b)   Neon-22 has how many neutrons?_____ Protons? _____

Electrons? _____

(a) 10; (b) 12, 10, 10 (assuming the atom to be electrically neutral)

2.   Since two atoms of the same element may contain different numbers of neutrons, nuclear physicists normally include the mass number when discussing an isotope. For example, the two isotopes of neon mentioned above are denoted $^{20}_{10}$Ne and $^{22}_{10}$Ne. The subscript denotes the atomic number and the superscript the mass number.* You can calculate the number of neutrons in an isotope by subtracting the atomic number from the mass number; that is, the number of neutrons is the difference between the two numbers.

$^{238}_{92}$U is sometimes written simply as U-238 and called uranium-238. How many

neutrons are in its nucleus? _____

146

RADIOACTIVITY

3.   A French physicist, Henri Becquerel, was experimenting with uranium in 1896 when he noticed that the uranium emitted invisible rays which were able to penetrate solid matter and expose photographic film. In honor of their discoverer, these rays were given the name Becquerel rays. Since the rays couldn't penetrate dense material very well, the uranium was enclosed in a lead block with a hole in one side to create a beam. The figure on the next page shows the experimental set-up which was used to determine the nature of

----

*The symbol and names of the elements are given in Table 5.1, on page 57.

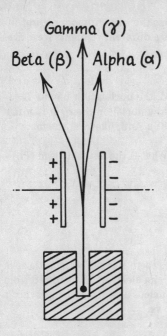

these rays. In the figure the rays are shown coming out of the hole in the lead block and passing between two sheets of metal which are electrically charged, one positive and the other negative. When this is done, the rays divide into three paths as shown.

Examine the figure and the deflection of the rays.

(a)   Which type of ray seems to have no electric

charge?_____

(b)   Which is positively charged? _____

(c)   Which is negatively charged? _____

— — — — — — — — — — — — — — — —

(a) gamma; (b) alpha; (b) beta

4.   The three types of radiation were originally given names of three letters of the Greek alphabet, α (alpha), β (beta), and γ (gamma). The alpha ray (or alpha particle, as it is called now) was found to be made up of two protons and two neutrons bound together. Since this is exactly the same as the nucleus of the helium-4 isotope, the alpha particle is symbolized in either of two ways: $_2^4$He or $_2^4\alpha$.

Beta particles were found to be electrons. And they are likewise denoted by either of two symbols: $_{-1}^0\beta$ or $_{-1}^0e$. The numbers here may take some explaining. The subscript in nuclear notation normally is the atomic number, or the number of protons (positive charge) in the nucleus. In this case, the particle has a negative charge, so it is given a subscript of negative one. The superscript is the mass number, the total number of nucleons (neutrons and protons). But an electron contains neither neutrons nor protons, and so it is given a superscript of zero.

The gamma ray is simply electromagnetic radiation. If you refer to the electromagnetic spectrum of Figure 17.3, page 190, you will see that the frequency range of gamma rays overlaps that of x-rays, but extends to higher frequencies also.

(a)   Write the symbol for an alpha particle. _____

(b)   Write the symbol for a beta particle. _____

(c)   Why is it gamma "ray," when the others are called "particles"? _____

_____

— — — — — — — — — — — — — — —

(a) α or $_2^4\alpha$ or $_2^4$He; (b) β or $_{-1}^0\beta$ or $_{-1}^0e$; (c) gamma is electromagnetic radiation, not particles

5.   When a nucleus as it exists is somehow unstable, it throws off radiation in order to become more stable. We call such an unstable isotope a radioactive isotope. For example, radium-226 is unstable in such a way that it emits an alpha particle. After this happens to

a single radium-226 nucleus, that nucleus no longer contains the same number of protons and neutrons that it did before the radiation, and thus it is a different element. We write this in formula form as follows:

$$^{226}_{88}Ra \rightarrow {}^{222}_{86}Rn + {}^{4}_{2}He$$

This formula tells us that once the alpha particle is emitted, the nucleus left behind is radon-222. Notice that the total of the atomic number of the nuclei on the right is equal to the atomic number of the original nucleus ($86 + 2 = 88$). And, likewise, the mass numbers total in the same way ($222 + 4 = 226$).

The isotope polonium-210 is also an alpha emitter. What is the resulting lead (Pb)

isotope? (Give the atomic number and the mass number.) _____

– – – – – – – – – – – – – – –

$^{206}_{84}Pb$ $({}^{210}_{84}Po \rightarrow {}^{206}_{82}Pb + {}^{4}_{2}He)$

6.   Some isotopes are unstable in a manner which causes them to emit beta particles in order to become more stable. But how can a beta particle—an electron—be emitted from the nucleus when none exists in the nucleus? Within the nucleus a neutron changes into a proton and an electron. We can write this reaction as follows:

$$n \rightarrow p + e$$

The electron is then thrown off from the nucleus. The isotope lead-209 is a beta emitter as shown in the following formula:

$$^{209}_{82}Pb \rightarrow {}^{209}_{83}Bi + {}^{0}_{-1}e$$

Notice that the atomic number of the resulting isotope is greater than that of the original isotope, since it has one more proton than before. The mass number does not change, however. The reaction balances just as do the ones which show alpha emission. Thus you should be able to complete the following beta emission formula. (You will have to look up the symbol of the isotope which results on pages 57 or 62.)

$$^{234}_{91}Pa \rightarrow \text{_____} + {}^{0}_{-1}e$$

– – – – – – – – – – – – – – –

$^{234}_{92}U$

7.   Following an alpha or beta emission by an isotope, the newly-formed nucleus, called the daughter nucleus, often has too much energy remaining. To rid itself of this energy, it emits a gamma ray. This does not change the basic nature of the isotope, but simply relieves it of excess energy. This is similar to the way that an atom with excess energy emits it in the form of a photon of light. (You may have studied this in Chapter 19.) Name the daughter nucleus in each of the cases below. (Find the symbol on page 57.)

(a) $^{214}_{84}Po$ emits an alpha particle. _____

(b) $^{14}_{6}C$ emits a beta particle. _____

(c) $^{238}_{92}U$ emits an alpha particle. _____

(d) $^{61}_{28}Ni$ emits a gamma ray. _____

(e)   What *does* happen to the Ni nucleus? _____

– – – – – – – – – – – – – – –

(a) $^{210}_{82}$Pb; (b) $^{14}_{7}$N; (c) $^{234}_{90}$Th; (d) $^{61}_{28}$Ni; (e) the nickel nucleus loses energy

HALF-LIFE

8.   Bismuth-210 atoms are beta radioactive, decaying (that is the term used by nuclear physicists) by beta emission to polonium-210. These nuclei do not all decay at once, however. Their decay is a chance occurrence, and a given nucleus may decay at any time. We can predict the probability of a bismuth-210 nucleus decaying during any time period, though, as a result of our observations of the decay of other Bi-210 nuclei. This prediction is based on the *half-life* of the isotope. The half-life of Bi-210 is 5 days, which means that in 5 days half of the original isotope will decay to Po-210. Thus if we start with 1 billion nuclei, we will have 500 million left after 5 days. After another 5 days we will have 250 million left. And after another 5 days 125 million will remain. A graph of the decay of Bi-210 would appear as shown in the figure at the left. Notice that as time passes, there is less and less bismuth left, since some is always changing to polonium. Thus, as each succeeding half-life passes there is less radiation.

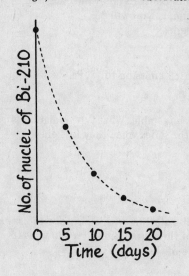

(a)   Will there be more beta emission between days 5 and 10 or days 20 and 25? _____

(b)   In the example above, how many bismuth nuclei will be left at the end of 20 days?

_____

– – – – – – – – – – – – – – –

(a) 5 and 10; (b) 62.5 million (which is half of what was listed for 15 days)

9.   Imagine that you have 1 billion pennies and you flip them into the air and give away all of those which come up heads. (These have "decayed.") You would expect about half to come up heads, or 500 million. Next you take the remaining coins and flip them. Again you give away the heads. If you imagine repeating this procedure over and over, you should see that a graph of the number left versus the number of times they were flipped would be similar to the half-life graph of the figure in frame 8. And the similarity is based upon the fact that both occurrences are governed by the rules of chance; we expect half of the coins to come up heads with each flip, and (from past experience) we expect half of the radioactive bismuth nuclei to decay in 5 days.

The half-lives of different radioactive substances are very different. Half-lives range from tiny fractions of a second up to billions of years.

Would an isotope with a half-life of 1 second or one with a half-life of one minute

be more radioactive? _____

– – – – – – – – – – – – – – –

1 second (assuming that you had the same number of radioactive nuclei to start with)

RADIOACTIVE SERIES

10.   You may have noticed that in the last example Bi-210 decayed to Po-210, but in the section on alpha decay Po-210 decayed into Pb-206. (You certainly are not expected to memorize these decay formulas, but if you check you will find that indeed these two were given above.) These two decays are the last two of a series of radioactive decays that starts with uranium-238.

Uranium-238 occurs in nature, and has a half-life of 4.5 billion years. If it weren't for this tremendously long half-life, this isotope wouldn't even exist on earth, for it is thought to have been formed as one of the products of the explosion of a star long before the formation of our solar system. The experiment which demonstrated that there were three types of radiation used uranium placed in a lead block. The reason that all three rays would be emitted by uranium ore is that uranium ore contains all of the daughter products of the radioactive series that starts with uranium, as you will see.

We'll follow a U-238 nucleus through part of its series:

(a) $^{238}_{92}U$ decays by alpha emission to _____Th.

(b)   This thorium isotope decays by _____ emission to $^{234}_{91}Pa$.

(c) $^{234}_{91}Pa$ beta decays to _____.

(d)   After 12 more decays, the result is lead-206, which is stable. What type radiation is emitted (besides α and β) by various isotopes in the series when they lose energy?

_____

— — — — — — — — — — — — — — —

(a) $^{234}_{90}Th$; (b) beta; (c) $^{234}_{92}U$; (d) gamma

(There are three other radioactive series, one of them also starting with an isotope of uranium. We will see in the next chapter that knowledge of these uranium series allows us to determine the age of rocks, and thereby a minimum age of the earth.)

SELF-TEST

The questions below will test your understanding of this chapter. Use a separate sheet of paper for your diagrams or calculations. Compare your answers with the answers provided following the test.

1.   In what way can one atom of an element differ fundamentally from another atom of the same element? _____

2.   What does the atomic number tell us about a nucleus? _____

3.   What does the mass number tell us about a nucleus? _____

4. $^{234}_{92}U$ has how many protons in its nucleus? _____ How many neutrons?

_____

5.   What is an alpha particle? _____

6.   What is a beta particle? _____

7.   What is a gamma ray? _____

8.   Thorium-229 decays by alpha emission. What is the product of the reaction? (Use the periodic table on page 62.) _____

9.   Actinium-225 is the daughter product of a beta emission. What was the isotope which emitted the beta particle? (Use the periodic table.) _____

10.   What change takes place in the nucleus when a gamma ray is emitted? _____

_____

11.   The half-life of actinium-225 is 10 days. If you started with 2 milligrams of pure Ac-225, how much would remain after 30 days? _____

12.   Why does a radioactive material which decays to a stable isotope become less radioactive as time passes? _____

_____

13.   Suppose that you start with some radioactive isotope and find that it emits only beta particles. After two weeks you reexamine it and find that now alpha particles are being emitted also. What would be your explanation? _____

_____

Answers to Self-Test

If your answers do not agree with those given below, review the frames indicated in parentheses before you go on to the next chapter.

1.   in the number of neutrons in their nuclei   (frame 1)

2.   the number of protons in the nucleus   (frame 1)

3.   the total of the neutrons and protons, or the number of nucleons   (frames 1–2)

4.   92 protons and 142 neutrons   (frame 2)

5.   the nucleus of a helium atom, two protons and two neutrons   (frame 4)

6.   an electron which has been emitted by a radioactive nucleus   (frame 4)

7.   electromagnetic radiation, of a frequency equal to that of x-rays or greater (frame 4)

8. $^{225}_{88}$Ra. Solution: $90 - 2 = 88$, the atomic number of the daughter. The periodic table shows this to be Ra (radium). $229 - 4 = 225$, the mass number.   (frame 5)

9. $^{225}_{88}$Ra ($^{225}_{88}$RA → $^{225}_{89}$Ac + $^{0}_{-1}e$)   (frame 6)

10.   The nucleus loses energy, but remains the same isotope it was.   (frame 7)

11.   0.25 milligrams.   Solution: 30 days = 3 half-lives. After 1 half-life: 1 milligram. After 2: $\frac{1}{2}$. After 3: $\frac{1}{4}$.   (frame 8)

12.   With each decay, there is one less nucleus left of the original radioactive isotope. Since there are fewer available for decay as time passes, fewer decay. (frames 8–9)

13.   Apparently the beta emitter decays to an isotope which is an alpha emitter. (frame 10)

CHAPTER TWENTY-SIX
Detection and Use of Radiation

Prerequisites: Chapters 5 and 25

Most people think of radiation as something harmful. And indeed, it is under many circumstances. But like most other things with which we come in contact in our lives, it has beneficial as well as harmful aspects. One of the dangers associated with nuclear radiation is that our bodies are unable to detect it directly.

Historically, the first detection of this radiation was based upon the fact that it is able to expose film. In this chapter, we will consider two other ways of detecting radiation: the Geiger counter and the cloud chamber. The property of radiation which makes possible these two methods (and most others) is the fact that radiation ionizes atoms as it passes by.

After completing this chapter, you will be able to:

- explain how alpha or beta particles cause ionization;
- specify how a Geiger counter measures radiation;
- specify how a cloud chamber can show the path of ionization;
- differentiate between absorption of beta or alpha particles and of gamma rays;
- name several biological effects of radiation;
- explain the purpose and cause of carbon-14 dating;
- write simple balanced nuclear reaction formulas;
- specify the use of neutrons in forming transuranic elements;
- describe an example of gauging with radiation;
- explain the purpose and method of uranium dating.

IONIZATION BY RADIATION

1. Imagine an alpha particle passing close to an atom as the particle moves through the air. The positively charged particle attracts the electrons circling the atom. This sudden force as the particle of radiation passes by can cause the electron to be jerked away from

the atom. It will probably not catch the alpha particle, however, because of the particle's great speed. The electron is simply separated from the atom so that the alpha particle leaves behind a free electron and a positively charged atom, called an ion.

A beta particle is a high speed electron, and as it passes by atoms, it exerts a repulsive force on other electrons. The result, however, is the same; a path of ionization is left behind the beta particle. The existence of ionization forms the basis of most methods of detection of radiation.

(a)   When an alpha particle passes through air, what two components are left in the path

of radiation? _____

(b)   What components are left in the path of beta radiation? _____

(c)   How does ionization caused by alpha and beta radiation differ? _____

_____

– – – – – – – – – – – – – – –

(a) free electrons and positive ions; (b) free electrons and positive ions; (c) only in direction of force that separates electrons from atoms (The results are the same.)

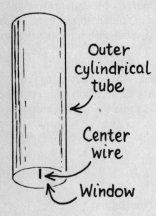

Outer cylindrical tube

Center wire

Window

THE GEIGER TUBE

2.   The figure at left is a diagram of a Geiger tube (or Geiger-Muller tube, or simply G-M tube). This device consists of a cylindrical metal tube insulated on the outside and with a wire down its center. The tube is connected to a source of high voltage, the center wire being made positive and the cylinder negative. In order to allow radiation to enter the tube without being absorbed by the metal walls, the end is made of a very thin material and serves as a "window" for the radiation.

Suppose an alpha or beta particle enters the window and travels through the tube. In its path it leaves ionization; but since the center wire is positive, the electrons are attracted toward it. As these particles are pulled in, they gain speed and each one knocks other electrons free. These electrons are pulled toward the wire and they, in turn, knock more electrons free. The result is that a great shower of electrons hits the center wire. This shower then passes on down to the voltage source, which contains a device to detect the small pulse of current formed by the electrons. It is this pulse which causes the clicking noise heard from Geiger counters in movies and on TV. In the laboratory, however, it is more likely that the pulse of electric current advances an electronic counter, or scalar, by one count. In this way, the total number of particles which pass through the tube in a given amount of time is counted.

(a)   What effect would you expect gamma radiation to have on a Geiger counter?

_____

(b)   What does the Geiger tube actually count, radiated particles or ionization?

_____

– – – – – – – – – – – – – – –

(a) none, because gamma rays have no charge and don't cause ionization; (b) ionization

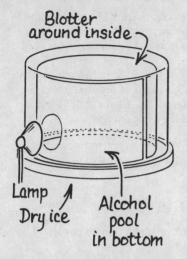

Blotter
around inside

Lamp
Dry ice

Alcohol
pool
in bottom

THE CLOUD CHAMBER

3.   The figure in this frame shows a diffusion cloud chamber. A small pool of alcohol placed in the bottom of the cylindrical chamber is soaked up by the blotter that covers the inside of the walls. Dry ice (solid carbon dioxide) below the chamber keeps the air at the bottom cooler than at the top. Alcohol evaporates from the top of the blotter and settles toward the bottom of the chamber as it cools. Because cool air cannot hold as much evaporated liquid as warmer air, the alcohol vapor is supersaturated near the bottom. In this state, the alcohol is ready to condense into droplets, but does not condense until it finds something to condense on. (This is the same principle as cloud seeding—the water in the cloud condenses on crystals dropped through the cloud.) If a radioactive source is placed in the cloud chamber (or if radiation is allowed to enter through a window in the side), the radiation leaves electrons and ions in its path, and these ions give the alcohol something on which to condense. Tiny alcohol droplets are formed along the path of the ray. The light shining in from the side allows us to see the droplets. In this way, one can actually see the path of the radiation and how far it penetrates through the air before being absorbed. If a cloud chamber is placed in a magnetic field, the beta particles and alpha particles will deflect in different directions due to the field. (See Chapter 16, frame 1 for the theory of this interaction.) It is by such experiments that radiation and its interactions are analyzed.

(a)   When a cloud chamber is used to detect radiation, what do we actually see to indicate radiation? _____

(b)   Which could you measure in a cloud chamber but not with a Geiger tube—the path of radiation, ionization presence, or gamma radiation? _____

— — — — — — — — — — — — — — — —

(a) condensed alcohol droplets; (b) path of radiation

ABSORPTION OF RADIATION

4.   We mentioned earlier that radiation causes electrons to be knocked from atoms. But this requires energy, and it is from the kinetic energy—energy of motion—of the moving particle that this energy comes. Thus the particle of radiation slows down as ionization proceeds.

The primary cause of the interaction between the radiation and the electrons of the matter through which it is passing is the electric charge of the particles. Thus the alpha particle, which has the most charge, causes the most ionization for each centimeter of matter through which it passes. This is one of the reasons that the range of alpha particles is only a few centimeters of air.*

_____

*There is another factor present here also—the much greater mass of the alpha particle. If you studied Chapter 2, you know that if an alpha particle and a beta particle have the same kinetic energy, the beta particle must have a far greater speed. (The mass of the alpha particle is nearly 7500 times the mass of the beta particle.) Thus the slow alpha particle exerts a force on an electron for a far greater time as it passes by. This means that it is much more likely to knock the electron free.

When an alpha particle stops, it picks up two free electrons from the surrounding matter and becomes an atom of helium. This is one of the ways by which the nature of the alpha particle was discovered—helium gas showed up in sealed containers with a radio-active source inside.

Beta particles have only half the charge of alpha particles and therefore are more penetrating. In a cloud chamber, they do not leave as dense a trail of droplets because they form fewer ions as they pass through each centimeter of air. A beta particle has a range hundreds of times greater than an alpha particle of the same energy.

(a)   Which has greater mass, an alpha particle or a beta particle? _____

(b)   If both have the same energy, which has greater speed, an alpha or a beta particle?

_____

(c)   Which forms helium atoms after stopping, alpha or beta particles? _____

(d)   If they have the same energy, which have greater range, alpha or beta particles?

_____

_ _ _ _ _ _ _ _ _ _ _ _ _ _ _

(a) alpha; (b) beta; (c) alpha; (d) beta

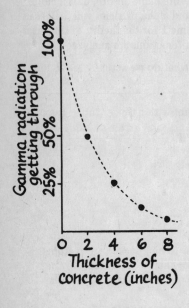

5.   The gamma ray has an entirely different character from the alpha and beta particle. Alpha and beta particles lose energy gradually in a series of interactions with electrons of the material through which they are passing. The gamma ray, however, normally loses all of its energy in one collision. And this collision is a chance event, depending primarily upon the density of electrons through which the gamma ray is passing.

When a beam of gamma rays enters a slab of concrete—commonly used as a shield against radiation—some of the rays are absorbed in the first millimeter of concrete while others penetrate a number of feet. The figure here is a graph of the absorption of a typical beam of gamma rays passing through concrete. Notice that it has the same shape as the half-life graph studied in the last chapter. In fact, one can talk of the half-thickness of concrete for gamma rays of a particular energy. As a rule of thumb, we can say that almost all gamma rays are absorbed by a couple of feet of concrete.

How does absorption of gamma rays differ from absorption of alpha or beta particles?

cles?_____

_ _ _ _ _ _ _ _ _ _ _ _ _ _ _

One collision takes all of the energy of one gamma ray.

6.   The absorption of radiation by matter forms the basis of the first industrial application of radiation we will discuss—*gauging*. Suppose that a manufacturer wishes to roll out

a sheet of steel of a certain thickness. He sets up a system in which, after the steel has passed through rollers, a beam of gamma rays passes through the steel and into a detector. The detector is set so that when a certain intensity of gamma radiation gets through to it, it does nothing. When fewer gamma rays hit the detector, a motor is activated to cause the rollers to make the steel thinner. And if too many rays hit it, it causes the steel to be rolled thicker. Such gauging techniques are used in the manufacture of many other products.

(a)   What feature of gamma radiation is used here?_____

(b)   Why do you suppose they use gamma rather than alpha radiation? _____

_____

— — — — — — — — — — — — — — — —

(a) the half-thickness can be predicted; (b) alpha particles won't penetrate the steel

RADIATION AND BIOLOGY

7.   In addition to its ionizing effect, radiation can tear molecules apart, and this destruction of molecules causes the death of cells in the body. Since the body naturally replaces cells, a certain amount of radiation can be tolerated, and indeed we are continually being irradiated by a number of sources of radiation in the environment. Some disagreement exists as to whether there is a threshold below which no permanent damage is done, but mankind has always coexisted with this background radiation.

A damaged cell cannot perform its normal function. If a number of cells in the same area are killed, the body suffers from the inability of these cells to function. Sometimes, rather than killing the cell, radiation causes it to reproduce at a great rate. The cell has become cancerous, and the growth—if not stopped—can end up killing the entire body. Even more frightening is damage to the reproductive cells, which determine the characteristics of any further offspring produced by that person. Radiation can thereby induce changes that can be passed on to the next generation. These are called mutations, and they are generally harmful or even lethal.

Name three possible effects of radiation on the body. _____

_____

— — — — — — — — — — — — — — — —

kill cells; cause uncontrolled growth of cells, or cancer; cause a change in reproductive cells

8.   Fast-growing cells are most susceptible to radiation damage. This leads to two conclusions: (1) pregnant women should avoid radiation, including x-rays, because the fast-growing baby might be injured; and (2) cancer (which is fast-growing) can be killed by radiation. Radiation therapy is one of our present treatments of cancer.

Radiation is also used medically for tracing the progress of various substances through the body. A radioactive isotope of an element which exists normally in the body is injected into, or fed to, a person. The radioactive isotope can be traced by following it with a radiation detector as it moves through the body. For example, after a patient has swallowed a small amount of radioactive iodine, the amount of iodine which concentrates in the thyroid gland can be determined.

(a)   Name two beneficial uses of radiation in medicine. _____

_____

(b)   Why are pregnant women advised to avoid x-rays?_____

_____

— — — — — — — — — — — — — — — —

(a) cancer treatment, tracing isotopes; (b) to avoid damage to the fast-growing baby

9.   Some research which uses radiation as a tool may simply involve placing a small
amount of radioactive element in air or water and tracing its path to discover the direction of air or water currents. Tracing is used in agriculture research to determine such
matters as fertilizer uptake into plants. The researcher includes in the fertilizer a radioactive isotope of an element which is normally found in fertilizer. The effectiveness of
various methods of fertilization can then be compared.
    Can you suggest a method of finding out how much of the isotope is taken up by

the plant?_____

— — — — — — — — — — — — — — — —

Hold a Geiger counter near the stem and leaves of the plant.

NUCLEAR DATING TECHNIQUES

10.   In the last chapter we talked about radioactive series, in which a radioactive isotope
decays to a stable isotope by a series of radiations. Uranium-235 and U-238 both occur
naturally, but when found they are never pure. They contain all of the elements in both
of the radioactive series starting with these two isotopes or uranium. At the end of each
of these series is a stable isotope of lead. By measuring the relative amounts of lead and
uranium in a uranium-bearing rock, one can determine the age of the rock.

(a)   The half-life of U-238 is about 4.5 billion years. In the oldest rocks found so far
      about half of the original U-238 has decayed. What, therefore, is the minimum age

      of the earth?_____

(b)   Are these rocks as radioactive now as they once were? _____

— — — — — — — — — — — — — — — —

(a) about 4.5 billion years; (b) no (about half as radioactive)

11.   Another nuclear dating method is used to determine the age of once-living matter.
A known percentage of all carbon in the atmosphere is the radioactive isotope carbon-14.
This isotope exists in a constant percentage because it is formed by cosmic rays, which are
assumed to have hit the earth at a relatively constant rate for tens of thousands of years.
All living matter takes carbon from the air, either directly (by plant matter) or indirectly
(by animals eating plants), and thus all living matter contains this same percentage of the
radioactive C-14. When the object dies, however, its carbon intake stops. The decay of
carbon-14 into stable nitrogen continues, however, with a half-life of 5800 years.
    To determine the age of an old bone, the researcher measures the percentage of
C-14. If the bone contains only half as much C-14 as living matter, for example, he knows

that the animal has been dead for about 5800 years. If it contains only one-fourth as much C-14, it died about 11,600 years ago. Using this method, a faily accurate age can be determined for once-living matter up to about 50,000 years old. The reason that there is a limitation to the age of materials analyzed by this method is that as the object gets older, there is less and less carbon-14 in the object. It therefore becomes very hard to detect.

(a)   Which would have stopped living more recently, an object with $\frac{1}{10}$ the C-14 or one with $\frac{1}{30}$ the C-14 of living matter? _____

(b)   Which of these could be dated using the C-14 method—uranium-bearing rocks, charred trees, or bones? _____

－－－－－－－－－－－－－－－－

(a) $\frac{1}{10}$; (b) charred trees and bones

NUCLEAR REACTIONS

12.   Nuclear physics is one of the largest branches of physics today. One of the areas of research within this branch of physics is the study of nuclear forces. Inside the nucleus we find protons, all of which have a positive charge and therefore repel one another. And yet the nucleus holds together. There must be some force, greater than the electrostatic force, holding these positive particles so close together. Researchers learn more about the nuclear force by causing nuclear reactions and observing the results. For example, if protons (obtained by ionizing hydrogen) are sent at a high speed through nitrogen gas, the following reaction is found to occur.

$$\mathrm{^{2}_{1}H + {}^{14}_{7}N \rightarrow {}^{12}_{6}C + {}^{4}_{2}He}$$

This reaction must balance just as did the nuclear decay reactions in the last chapter, with the mass numbers and the atomic numbers being equal on each side.

The reaction below produces hydrogen and another element. Use the periodic chart as needed, and complete the reaction formula.

$$\mathrm{^{4}_{2}He + {}^{28}_{14}Si \rightarrow {}^{1}\_H + \underline{\hspace{3cm}}}$$

－－－－－－－－－－－－－－－

$\mathrm{^{1}_{1}H + {}^{31}_{15}P}$ (Phosphorus is element 15.)

13.   From the energies of the particles involved in reactions such as those in frame 12, physicists are able to learn some of the characteristics of nuclear forces.

Of particular interest are reactions which yield neutrons. For example:

$$\mathrm{^{4}_{2}He + {}^{9}_{4}Be \rightarrow {}^{12}_{6}C + {}^{1}_{0}n}$$

This reaction produces a neutron (n), which makes a particularly good "bullet" to shoot at other nuclei. It does not experience the electrostatic repulsive force which is experienced by protons and alpha particles, and so it is more likely to penetrate large nuclei and cause a further nuclear reaction. It is by use of neutron bombardment that synthetic elements were formed.

(a)   Would a large nucleus have a large or small charge? _____

(b)   Which would be repelled most by a nucleus—an alpha particle, a proton, or a

neutron? _____

– – – – – – – – – – – – – – – –

(a) large; (b) alpha (its charge is $+2$)

14.   When a beam of neutrons strikes uranium, the following reaction occurs,

$$^{1}_{0}n + {}^{238}_{92}U \rightarrow {}^{239}_{92}U$$

The story does not stop here, however, for U-239 is beta-radioactive.

$$^{239}_{92}U \rightarrow {}^{239}_{93}Np + {}^{0}_{-1}e$$

And neptunium-239 is also beta-radioactive, producing as a daughter nucleus element number 94, plutonium. Neptunium and plutonium were the first two synthetic elements. Since they are beyond uranium in the periodic chart, they are called *transuranic* elements.

By bombarding the heavy elements with various particles, all transuranic elements up to element number 105 have been formed. They were named by the research groups who discovered them and include such names as americium, californium, and einsteinium (the origins of which should be obvious to you).

What type bombardment produced the first transuranic elements? _____

– – – – – – – – – – – – – – – –

neutron

SELF-TEST

The questions below will test your understanding of this chapter. Use a separate sheet of paper for your diagrams or calculations. Compare your answers with the answers provided following the test.

1.   What happens to atoms of the material through which alpha and beta particles pass?

_____

_____

2.   What advantages does the cloud chamber have over the Geiger tube? _____

_____

3.   Why does the beta particle have a greater range than the alpha particle? _____

_____

4.   How might gauging be used in an industrial process? _____

_____

_____

5.   What nuclear method is used in establishing the age of ancient rocks?

_____

6.   When an alpha particle strikes the nucleus of a $^{14}_{7}N$ atom, two products result. One is a proton. What is the other? (Use the periodic table.)_____

7.   Why is the neutron especially valuable as a particle with which to initiate a nuclear reaction? _____

_____

8.   What is meant by the term "transuranic element"? _____

_____

9.   What type radiation requires us to use thick concrete for shielding?_____

_____

Answers to Self-Test

If your answers do not agree with those given below, review the frames indicated in parentheses before you go on to the next chapter.

1.   Electrons are jarred from the atoms. Ions and free electrons are left in the particle's path. (In addition, the atom may be broken free from its molecule.)   (frames 1, 7)

2.   In the cloud chamber, one can see the path taken by the ionizing radiation. (frame 3)

3.   The beta particle has less charge. In addition, because its mass is less it is moving faster.   (frame 4)

4.   Gamma radiation is passed through sheets of material and then monitored. Depending upon the amount of radiation getting through, the sheets are rolled thicker or thinner.   (frame 6)

5.   uranium dating   (frame 10)

6. $^{17}_{8}O$. Solution: $^{4}_{2}He + {}^{14}_{7}N \rightarrow {}^{1}_{1}H + {}^{17}_{8}O$ (frame 12)

7.   The neutron has no electric charge, so it experiences no repulsive force from the positive nucleus.   (frame 13)

8.   A transuranic element is one of the elements with an atomic number greater than that of uranium.   (frame 14)

9.   gamma rays, since they penetrate much farther than the other two radiations (frame 5)

CHAPTER TWENTY-SEVEN
Fission, Fusion, and Energy

Prerequisites: Chapters 5, 25, and 26

In this chapter we will examine the greatest source of energy presently known to man—nuclear energy.* The history of the development of nuclear energy (including the nuclear bomb) is very interesting, but we will limit ourselves here to the physics of nuclear energy. You are encouraged to delve into the historical aspects on your own.

Studying this energetic chapter will enable you to:

- specify the source of energy in nuclear reactions;

- give an example of a fission reaction;

- specify what is necessary to form a nuclear chain reaction;

- specify the effect of a chain reaction on mass and energy;

- compare enriched uranium to natural uranium;

- explain the effect of a moderator on a fission chain reaction;

- specify the meaning of "critical mass" of uranium;

- identify the components of a fission power reactor as fuel elements, control rods, and water, and state the function of each;

- specify the role of critical mass, moderator, enrichment, and lead covering in a fission bomb;

- identify the current status of fusion reactors;

- specify the importance of the multiplying factor in a chain reaction.

---

*The expression "nuclear energy" includes solar energy; the source of the energy of stars is nuclear fusion.

ENERGY IN NUCLEAR REACTIONS

1.   When an alpha or beta particle is released from a radioactive nucleus, it comes out of the nucleus at a great speed. Thus it has kinetic energy. Where did this energy come from? The answer lies in Einstein's famous equation: $E = mc^2$. (Here E is energy, m is mass, and c is the speed of light.) This equation can be used to calculate how much energy is produced when a certain amount of mass goes out of existence. This literally means "goes out of existence," and not simply "disappears." In alpha decay, the total mass of the alpha particle and daughter nucleus is less than the mass of the parent nucleus. And the amount of missing mass is just enough to account, by Einstein's equation, for the kinetic energy gained by the products of the decay.

(a)   How is energy produced in nuclear decay? _____

(b)   What does $E = mc^2$ have to do with nuclear reactions? _____

_____

– – – – – – – – – – – – – – – –

(a) by conversion of mass to energy; (b) it gives the amount of energy generated by a given amount of lost mass

NUCLEAR FISSION

2.   In frame 13 of Chapter 26 we saw that a neutron is an especially good projectile to fire at a large nucleus. One of the isotopes which exists in all natural uranium in small amounts is U-235. When a neutron is absorbed by the U-235 nucleus, a number of different reactions may occur. Here are three reactions of the many that are known.

$$_0^1 n + {}_{92}^{235}U \rightarrow {}_{56}^{142}Ba + {}_{36}^{91}Kr + 3\,_0^1 n$$

$$_0^1 n + {}_{92}^{235}U \rightarrow {}_{57}^{139}La + {}_{42}^{95}Mo + 2\,_0^1 n$$

$$_0^1 n + {}_{92}^{235}U \rightarrow {}_{51}^{133}Sb + {}_{39}^{95}Yt + {}_2^4He + 4\,_0^1 n$$

These are called *fission reactions* because the uranium nucleus fissions, or breaks apart. Only six of the more than 200 fission products that result from the fission of U-235 are shown. You are certainly not expected to memorize these; they are presented as examples.

Two important points must be made about the reactions, however. First, the total mass of the products is less than the total mass of the neutron and U-235 nucleus which went into the reaction, indicating that energy was released. This energy shows up as kinetic energy when the products of the reaction separate at great speed (and in energy of gamma rays which are released but were not shown in the formulas above). Second, neutrons were released as products of the reaction. In fact, though only one neutron started the reaction, two to four neutrons are released by each fission.

(a)   In the reactions above, would you expect the neutrons released to travel at a high

or a low speed? _____

(b)   Is the total mass of the particles and isotopes on the right greater or less than on

the left? _____

(c)   What is the total of the mass numbers on the left side of the first formula? _____

On the right? _____

(d)   Why are these called fission reactions? _____

_ _ _ _ _ _ _ _ _ _ _ _ _ _ _

(a) very high speed; (b) less; (c) 236, 236; (d) the original nuclei broke apart

3.   Suppose that a neutron hits a U-235 nucleus and three neutrons are released. Suppose further that each of these hits another U-235 nucleus, they fission, and three more
neutrons are released by each. And that these nine neutrons cause nine more fissions,
releasing three neutrons each. Then these 27 grow to 81. Then 243, 729, 2187, and so on.
When you consider that the time involved in each of these fissions is around a millisecond,
you realize that this chain reaction process is extremely quick.

Because of the loss of mass that occurs with each fission, the fission products separate at great speeds. This great speed, or kinetic energy of the particles, shows up as *heat*.
The uranium explodes with great energy!

Refer to the three fission reaction formulas in the preceding frame.

(a)   Which would you expect to contribute most to the growth of the chain reaction?

_____

(b)   Where does the energy come from for the great increase in heat? _____

_ _ _ _ _ _ _ _ _ _ _ _ _ _ _

(a) third one, because it releases 4 neutrons; (b) the loss of mass

4.   Based on the chain reaction, it might seem very simple to build a nuclear fission
bomb. Fortunately, there are some complicating factors. First, only about 0.7% of
natural uranium is U-235. U-238, which makes up almost all of the rest, does not fission
when struck by a neutron. It emits a beta particle to become the transuranic element
neptunium. Since no neutrons are produced, there is no chance of a chain reaction from
U-238 absorbing a neutron. Because the many U-238 nuclei absorb neutrons without
releasing any, the neutrons produced by a single fission of U-235 will probably be absorbed by U-238 nuclei and no chain reaction will occur. To make a chain reaction occur,
one must somehow arrange that there are more U-235 nuclei available to be struck by
neutrons.

Even when present, however, U-235 nuclei tend to absorb only slow-moving neutrons,
and the nonfissioning U-238 nuclei tend to absorb only fast neutrons. Since neutrons
emitted in fission are fast-moving, they are unlikely to cause further fission.

If a growing chain reaction is to take place, an average of greater than one neutron
released by each fission must produce another fission. Because of the effects discussed
above, far fewer than one of the $2\frac{1}{2}$ neutrons released on the average by fission strikes a
U-235 nucleus. Thus the fissions that constantly take place in uranium found in the earth
do not result in nuclear chain reactions—uranium mines do not explode. Our problem, if
we are to produce a chain reaction, is to somehow increase the multiplying factor—the
average number of neutrons from one fission which cause another fission.

(a)   Give two factors that decrease the probability of a chain reaction. _____

_____

(b)   Which of these multiplying factors will enable a chain reaction—0.99, 1.01, 1.1?

_____

— — — — — — — — — — — — — — — —

(a) U-235 is rare, and U-235 absorbs slow neutrons rather than fast ones; (b) 1.01 or 1.1

5.   One method of increasing the multiplying factor for chain reactions is using *enriched uranium*, which has a greater-than-normal fraction of U-235. Some 99.3% of naturally occurring uranium is the isotope U-238. If somehow the percentage of U-235 can be increased, the chances of a chain reaction will be increased. Since these two nuclei are isotopes of the same element, their chemical properties are extremely similar. Separation of the two isotopes is very difficult (and in fact, this was the major task facing the builders of the first nuclear bomb).

    We will not go into detail on how the separation may be achieved, but will simply point out that if a U-235 nucleus and a U-238 nucleus have the same kinetic energy, the one with the smaller mass will have the greater speed. (See Chapter 2, frame 12.) This allows us to separate the two nuclei, but the process is still a very difficult and expensive one.

(a)   What is enriched uranium? _____

(b)   If a U-238 nucleus and a U-235 nucleus have the same kinetic energy, which will

    have the greater speed? _____

— — — — — — — — — — — — — — — —

(a) higher than normal percentage of U-235; (b) U-235, because it has the smaller mass

6.   Even if we were able to isolate pure U-235, a fission reaction would occur only if enough of the isotope were present. Nuclei occupy a small percentage of the volume of a material, and a neutron normally passes by countless nuclei before it happens to hit one. Thus if the amount of uranium present is small, the neutron is likely to pass out of the uranium entirely before it can react with a nucleus.

    The minimum amount of uranium which will support a growing chain reaction is called the *critical mass* of the uranium. The critical mass amount depends on several factors: (1) the degree of enrichment of the uranium; (2) the shape in which the uranium is arranged—a spherical shape has a minimum surface from which neutrons can escape; and (3) whether a moderator—to be discussed in the next frame—is present to slow down neutrons.

(a)   Would you expect the critical mass to be greater or less in a more enriched uranium

    sample? _____

(b)   Would you expect a sphere or a cube of uranium to have a larger critical mass?

    _____

— — — — — — — — — — — — — — — —

(a) less, because there is a higher percentage of U-235 in enriched uranium; (b) cube, because there would be a greater probability that the neutrons would escape

7.   Uranium-235 is more likely to absorb a neutron and fission when struck by a slow neutron. How might we slow down neutrons? Well, suppose a golf ball is hit through a showroom of bowling balls. When the golf ball hits a bowling ball, the bowling ball will rebound hardly at all; the golf ball will bounce from the bowling ball with nearly the same speed that it hit. But now suppose that this same golf ball is hit through a room of baseballs. When it hits a baseball, the baseball will rebound more than the bowling ball did. Thus it takes some energy from the golf ball, which will then rebound with considerably less speed. After a number of collisions with baseballs, the golf ball will be moving at a slow speed.

The same principle is used to slow down neutrons. If a neutron bounces from a large nucleus, the neutron loses very little of its speed. But if it bounces from a nucleus which has a mass comparable to its own, it loses a significant amount of its energy. Thus to slow down, to *moderate*, neutrons, we surround the fissioning material with a material which has nuclei of little mass. The moderator used in most nuclear reactors today is water, each molecule of which has two hydrogen atoms. Neutrons passing through water bounce around as they hit hydrogen nuclei, and they are slowed down. Thus they become more likely to initiate fission of a U-235 nucleus.

(a)   Why is the uranium in nuclear reactors surrounded by water? _____

_____

(b)   Water also contains oxygen atoms. Does a neutron lose more energy in collision

with oxygen or hydrogen nuclei? _____

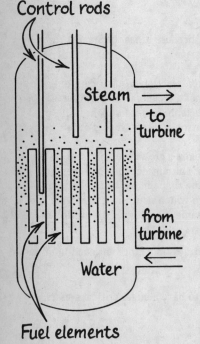

Control rods

Steam →

to turbine

from turbine

←

Water

Fuel elements

Figure 27.1

(c)   How does the mass of a hydrogen nucleus compare to the mass of a neutron? _____

– – – – – – – – – – – – – – – –

(a) to moderate the neutron speed; (b) hydrogen; (c) very similar

THE POWER REACTOR

8.   In a conventional fossil fuel power plant, coal, oil, or natural gas is burned to heat water to run turbines to produce electricity. The nuclear plant differs primarily in that the fission reaction produces the necessary heat. Figure 27.1 illustrates the reactor portion of a boiling water nuclear power reactor. Within the core of the reactor are fuel elements containing the enriched uranium fuel. The water flows between the fuel elements and serves a dual function: it acts as a moderator, and it is boiled by the heat of the reaction so that it is able to drive the turbine which is connected to the electric generator.

(a)   What drives the turbines from the nuclear

reactor here? _____

_____

(b)   What element slows down the neutrons? _____

(c)   Where is the enriched uranium? _____

— — — — — — — — — — — — — —

(a) steam from water boiled by nuclear energy; (b) water, or actually hydrogen nuclei; (c) in the fuel elements

9.   If the reactor in Figure 27.1 were operated with a critical mass of fuel and there were no way to limit the reaction, the heat produced could cause enough pressure to explode the walls of the reactor. It would not explode like a nuclear bomb, however, because once too much water boils away, the water would no longer be between the fuel elements to serve as a moderator. This serves as a safety feature, since it would stop the chain reaction and automatically shut down the reaction.

Even though a nuclear explosion is extremely unlikely, the bursting of the walls would still release radioactive material to the surroundings. In order to control the reaction, control rods—made of a material which absorbs neutrons—are placed through the core of the reactor. When the rods are in place in the reactor they absorb enough neutrons that the multiplying factor is less than 1. To start the chain reaction, the rods are slowly removed, allowing more and more neutrons to initiate fission. At some point the chain reaction becomes self-sustaining, the water boils, and electric power flows from the generator.

(a)   What is the function of the control rods in a power reactor? _____

_____

(b)   How likely is a nuclear explosion in a power reactor? _____

— — — — — — — — — — — — — —

(a) to absorb neutrons and thereby control the reaction; (b) extremely unlikely

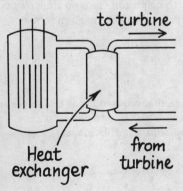

to turbine

from turbine

Heat exchanger

10.   About half of the reactors being built in the United States are of the type pictured in Figure 27.1 The other half are pressurized water reactors. The figure here shows that in these reactors there is another step involved. The water in the reactor core is kept at such a high pressure that it is unable to boil although it is heated to a great temperature. This water then goes through a heat exchanger, where it is used to boil other water which goes to the turbine to turn the electric generator.

What is used as a moderator in the pressurized reactor? _____

— — — — — — — — — — — — — —

water

THE BOMB

11.   As fuel, the power reactor uses partially enriched uranium, but a moderator is still necessary for a chain reaction to occur. In a bomb, we can't have the extra size taken up by a moderator. If enough highly enriched uranium is located in one place, a growing chain reaction will start even without a moderator. The uranium in a fission bomb is kept in two parts, each less than one critical mass. When the time comes to explode the bomb, the two masses of uranium are forced together and the reaction starts. In order to prevent the uranium from exploding apart before the reaction has grown to its maximum, the uranium is surrounded by lead. The lead is very dense and thus has enough inertia to hold the uranium in place for an extra tiny fraction of a second. Fission after fission occurs with tremendous heat build-up, until the powerful destructive force of the "atomic bomb" is released. ("Atomic bomb" is actually a misnomer; a better name is "nuclear bomb.")

(a)   How is the critical mass of uranium considered in building a bomb? _____

_____

(b)   Which is more of a factor in a nuclear bomb, enrichment or moderation?

_____

(c)   What is the function of the lead covering of a nuclear bomb? _____

_____

— — — — — — — — — — — — — —

(a) each piece must be *less* than the critical mass, but the total mass must be more than critical; (b) enrichment; (c) to keep the mass together until maximum reaction occurs

FUSION

12.   The energy given off when small nuclei combine is the basis of the *fusion reaction.* The fusion reaction used in the fusion bomb—normally called the hydrogen bomb—uses an isotope of hydrogen that contains a neutron along with its proton. Two possible results of the fusion of two of these deuterium (or "heavy" hydrogen) nuclei are given by the following formulas.

$$^2_1H + {}^2_1H \rightarrow {}^3_2He + {}^1_0n + \text{energy}$$
$$^2_1H + {}^2_1H \rightarrow {}^3_1H + {}^1_1H + \text{energy}$$

The energy released per pound of fuel is even greater than that released in the fission reaction. In addition, while the fission bomb has a size limit of less than two critical masses, the hydrogen bomb has no theoretical limit. For these reasons, the H-bomb has become the world's major nuclear weapon.

(a)   Why do you suppose hydrogen-2 is called "heavy" hydrogen?_____

_____

(b)   Which type bomb, fission or fusion, can pack more energy? _____

(c)   What is the fuel in the fusion bomb? _____

— — — — — — — — — — — — — —

(a) it has a neutron as well as a proton in its nucleus, so it is twice as massive; (b) fusion; (c) hydrogen-2

14.   The fusion reaction does not occur easily. Since both of the reacting nuclei are positively charged, they repel one another, and extra energy must be supplied to get them together. In the hydrogen bomb, this energy is provided by a fission bomb, which is actually exploded to ignite the hydrogen bomb.

Though man did not discover the fusion reaction until the middle of this century, fusion is the energy of the stars. Within our sun, the two reactions shown in frame 13 occur along with further reactions such as this.

$$_1^2H + {}_1^3H \rightarrow {}_2^4He + {}_0^1n + energy$$

More than 4 million tons of mass are being converted to energy each second in the sun, and this energy is radiated away to shine on our tiny earth and countless other planets throughout the universe.

(a)   Is the total mass of the nucleus (and neutron) on the right side of the formula in

this frame greater or less than the total mass on the left? _____

(b)   What element is produced by the reaction above? _____

(c)   How do we provide energy to start a fusion bomb reaction? _____

_____

– – – – – – – – – – – – – – – –

(a) less; (b) helium; (c) by exploding a fission bomb

THE FUSION REACTOR

15.   At present, fission reactors are being built to supply the world with needed energy. These reactors are not the ultimate solution to our energy problem, however. Besides the radiation pollution problems of fission power, the amount of uranium fuel in the world is limited. With fusion power, though, there is almost no limit to the availability of deuterium fuel. This isotope makes up a small percentage of all naturally occurring hydrogen. In every 30 kilograms (about 65 pounds) of water is one gram of deuterium. This does not seem like much fuel until one considers that if each nucleus of the gram of deuterium could be made to fuse, the energy emitted would be equivalent to the burning of 16 tons of coal!

We have used fusion energy in bombs for a quarter century. We have not been able to control it for use as a power source, however, since we cannot generate the tremendous temperature needed to ignite the fusion reaction. At the temperature needed, matter does not exist in the atomic state, but electrons are stripped from their nuclei, so no material container is capable of holding the reacting fuel. Research underway now concentrates on controlling the reaction by the use of magnetic fields. The problem is the greatest ever faced by the scientific and technological community. Whether fusion power becomes our ultimate energy source depends upon man's desire and ingenuity.

(a)   Why don't we have fusion reactors for energy production? _____

_____

(b)   Why is fusion seen as a very attractive energy alternative for the future? _____

_____

(c)   What are two disadvantages of fission as a source of energy? _____
_____

– – – – – – – – – – – – – – – – –

(a) tremendously high temperatures are needed for the reaction; (b) there is an essentially unlimited fuel supply; (c) radiation pollution and limited supply of fuel

How mankind uses the energy sources developed by science and technology depends upon our wisdom and our moral values. We can use them for destruction, or we can use them for the benefit of life on our planet. The decision is being made right now. And whether we like it or not, each of us has a voice in the decision.

SELF-TEST

The questions below will test your understanding of this chapter. Use a separate sheet of paper for your diagrams or calculations. Compare your answers with the answers provided following the test.

1.   How does Einstein's theory of mass-energy equivalence apply to the fission reaction?
_____
_____

2.   Which isotope of uranium is fissionable? _____

3.   What is it about fission reactions which makes a sustaining chain reaction possible?
_____
_____

4.   What is meant by the multiplying factor in a fission reaction? How great must the multiplying factor be to produce a self-sustaining reaction? _____
_____
_____

5.   What is enriched uranium? _____

6.   What is meant by a critical mass of uranium? _____
_____

7.   If one uses a moderator which is not very effective, will the critical mass be greater or less (all other factors remaining the same)? _____

8.   What type element is used as a moderator? Why? _____
_____

9.   What two functions does the water serve in a nuclear power reactor? _____
      _____

10.   What is the function of control rods in a reactor? _____
      _____

11.   Why is there such difficulty in controlling the fusion reaction? _____
      _____

Answers to Self-Test

If your answers do not agree with those given below, review the frames indicated in parentheses.

1.   The mass of the nuclei entering the reaction is greater than the mass of the products.
      The lost mass becomes energy.   (frame 2)

2.   U-235   (frames 2, 4)

3.   A neutron initiates the reaction and neutrons are produced by the reaction. These
      start new reactions.   (frame 3)

4.   It is the average number of neutrons released in one fission which produce another
      fission. For a reaction to be self-sustaining, it must be 1 or greater. (For the reac-
      tion to *grow*, it must be greater than 1.)   (frame 4)

5.   Enriched uranium contains a greater-than-normal percentage of U-235.   (frame 5)

6.   A critical mass of uranium is the minimum mass which, under given circumstances,
      will sustain a chain reaction.   (frame 6)

7.   greater   (frames 6–7)

8.   A light element is the best moderator because when the neutron strikes a light
      nucleus, it loses more of its energy than when it strikes a heavy nucleus.
      (frame 7)

9.   It serves as a moderator. It removes heat from the core of the reactor and transfers
      the energy to the turbine.   (frame 8)

10.   They absorb neutrons and thereby serve as a method of adjusting the multiplying
      factor. Thus they can be used to increase or decrease the reaction.   (frame 9)

11.   Fusion requires a very great temperature. Matter cannot exist as we know it at such
      a temperature.   (frame 15)

APPENDIX ONE

Scientific Notation: Powers of Ten

Science deals with such large and small numbers that it is often inconvenient to write them in the standard form. The form used in science is called *scientific notation*, or powers of ten notation. Both large and small numbers can be expressed as a number between 1 and 10, times 10 raised to some power.

METHOD OF WRITING LARGE NUMBERS

Look at the following powers of ten.

$$10^1 = 10$$
$$10^2 = 100$$
$$10^3 = 1000$$
$$10^4 = 10,000$$
$$10^{23} = 1 \text{ followed by 23 zeros}$$

The number 2,553,000 is $2553 \cdot 10^3$, or in scientific notation, it is $2.553 \cdot 10^6$. To check that this latter number is the same as the original, move the decimal place six places to the right, as shown below.

$$2\ 5\ 5\ 3\ 0\ 0\ 0\ .$$

1, 2, 3, 4, 5, 6 places to the right

Here are a few more examples. You can check to make sure they are equal.

$$473,000 = 4.73 \cdot 10^5$$
$$760,000,000 = 7.6 \cdot 10^8$$
$$4.35 \cdot 10^3 = 4350$$
$$5.747 \cdot 10^2 = 574.7$$

METHOD OF WRITING SMALL NUMBERS

The power of ten tells you how far to move the decimal point to the right. The system works with negative powers also, but in this case the decimal point must be moved to the left. For example:

$$10^{-2} = .01$$

(We have taken the number 1.0 and moved the decimal point two places to the left.)

$$10^{-3} = .001$$
$$10^{-5} = .00001$$

More complicated numbers follow the same pattern.

$$4.6 \cdot 10^{-4} = .00046$$

The decimal point here has been moved four places to the left, the zeros being supplied. Compare these examples.

$$3.24 \cdot 10^{-5} = .0000324$$
$$.00067 = 6.7 \cdot 10^{-4}$$
$$.0000005 = 5 \cdot 10^{-7}$$

Optional   CALCULATIONS WITH SCIENTIFIC NOTATION

To multiply and divide in scientific notation, follow two simple rules: (1) Handle the first part of the number and the power of ten part separately; and (2) multiply powers of ten by adding the powers, and divide by subtracting the powers. Some examples will illustrate.

$$(4 \cdot 10^2)(6 \cdot 10^3) = 24 \cdot 10^5 = 2.4 \cdot 10^6$$

Notice that although the interim answer was correct, it was changed to conform to standard scientific notation.

$$(1.3 \cdot 10^6)(2.2 \cdot 10^{-2}) = 2.86 \cdot 10^4$$

Here the 6 and the −2 were added, but since the second number was negative, this amounts to subtracting it from the first.

$$(1.33 \cdot 10^{-18})(3 \cdot 10^5) = 3.99 \cdot 10^{-13}$$
$$(1.33 \cdot 10^{-6})(3 \cdot 10^4) = 3.99 \cdot 10^{-2}$$
$$(1.33 \cdot 10^{-5})(3 \cdot 10^{-3}) = 3.99 \cdot 10^{-8}$$

Now for some division.

$$\frac{6 \cdot 10^8}{2 \cdot 10^3} = 3 \cdot 10^5$$

Here, 6 was divided by 2, and 10^8 was divided by 10^3 by subtracting the 3 from the 8.

$$\frac{2.6 \cdot 10^9}{2 \cdot 10^{-4}} = 1.3 \cdot 10^{13}$$

This results because $9 - (-4) = 9 + 4 = 13$. Now a combination.

$$\frac{(2 \cdot 10^5)(8 \cdot 10^{-2})}{4 \cdot 10^3} = \frac{16 \cdot 10^3}{4 \cdot 10^3} = 4$$

It doesn't matter whether the multiplication or the division is done first, but I chose to multiply first. You try it the other way for practice.

SELF-TEST

1. The brightest star in our sky is Sirius, which is 51,700,000,000,000 miles away. Write this number in scientific notation. _____

2. The speed of light is about $3 \cdot 10^8$ meters/second. Write this number in "standard" notation. _____

3. The mass of a proton is $1.67 \cdot 10^{-27}$ kilograms. Write this number without using powers of ten. _____

4. Write the number 0.000455 in scientific notation. _____

5. *Optional.* The mass of the proton is $1.67 \cdot 10^{-27}$ kilograms. What is the mass of a billion (10^9) protons?_____

6. *Optional.* The mass of the electron is $9.1 \cdot 10^{-31}$ kilograms. How many electrons are needed to have the same mass as one proton? (See the last question for the proton mass.) _____

7. *Optional.* The earth gets as far from the sun as $9.5 \cdot 10^7$ miles and Mars gets as close as $1.28 \cdot 10^8$ miles. How close does the earth get to Mars? (Hint: Subtract the two numbers.) _____

Answers to Self-Test

1. $5.17 \cdot 10^{13}$ miles

2. 300,000,000 meters/second

3. 0.000,000,000,000,000,000,000,000,001,67 kilograms (The decimal has been moved 27 places, requiring 26 zeros between the decimal point and the first non-zero numeral.)

4. $4.55 \cdot 10^{-4}$

5. $1.67 \cdot 10^{-18}$ kilograms

6. $1.84 \cdot 10^3$ or 1840

7. $3.3 \cdot 10^7$ miles.   Solution: The trick here is to write the numbers with the same power of ten. Using the power 7:

$$1.28 \cdot 10^8 = 12.8 \cdot 10^7$$

Now subtract:

$$
\begin{array}{r}
12.8 \cdot 10^7 \\
-9.5 \cdot 10^7 \\
\hline
3.3 \cdot 10^7
\end{array}
$$

APPENDIX TWO

The Metric System

In a few years, new books will not include an appendix on the metric system (though they may have one on the British system). The British system that we use will soon be a thing of the past for two reasons: First, it is desirable to have a single worldwide system, and second, the metric system is basically simpler than the British system.

LENGTH

The basic unit of length in SI units* is the meter. (A meter is 39.37 inches, a little longer than a yard.) The meter is divided into 100 centimeters, or into 1000 millimeters. Thus there are 10 millimeters in a centimeter. The only other unit of length which will be used in everyday language is the kilometer, which is 1000 meters. These lengths, as well as a number of other that are useful in science, are given below:

$$1 \text{ kilometer (km)} = 1000 \text{ meters (m)}$$
$$100 \text{ centimeters (cm)} = 1 \text{ m}$$
$$1000 \text{ millimeters (mm)} = 1 \text{ m}$$
$$10^6 \text{ micrometers } (\mu\text{m}) = 1 \text{ m}$$
$$10^9 \text{ nanometers (nm)} = 1 \text{ m}$$

Notice that all units are related to all others by a multiple of 10, and usually by a factor of 1000. This is why the metric system is basically simpler than the British system; there are no factors of 12, 3, or 5280.

In the metric system, it is easy to change a measurement made in one unit of length to another unit of length. For example, 52 centimeters is 0.52 meters. (Notice that this is the same as saying that 52¢ = $.52.) And 52 cm is 520 mm, or 520,000 μm, and so on.

---

*SI is the abbreviation for the French words for "International System." This is the metric system.

MASS

In the British system we express weight in pounds. As explained in Chapter 1, there is a distinction between weight and mass, mass being the more fundamental quantity. And in the metric system we will speak of a person's mass rather than weight.

The basic SI unit of mass is the kilogram (kg). One kilogram weighs about 2.2 pounds on earth. Knowing what the prefix "kilo" means, you know that a kilogram is 1000 grams. Although we do not speak of centigrams, the units of milligram (1000 mg = 1 g) and microgram (10^6 μg = 1 g) are used in science.

APPENDIX THREE
The Sine Table

Angle	Sine	Angle	Sine	Angle	Sine
0°	.000	31°	.515	61°	.875
1°	.018	32°	.530	62°	.883
2°	.035	33°	.545	63°	.891
3°	.052	34°	.559	64°	.899
4°	.070	35°	.574	65°	.906
5°	.087	36°	.588	66°	.914
6°	.105	37°	.602	67°	.921
7°	.123	38°	.616	68°	.927
8°	.139	39°	.629	69°	.937
9°	.156	40°	.643	70°	.940
10°	.174	41°	.656	71°	.946
11°	.191	42°	.669	72°	.951
12°	.208	43°	.682	73°	.956
13°	.225	44°	.695	74°	.961
14°	.242	45°	.707	75°	.966
15°	.259	46°	.719	76°	.970
16°	.276	47°	.731	77°	.974
17°	.292	48°	.743	78°	.978
18°	.309	49°	.755	79°	.982
19°	.326	50°	.766	80°	.985
20°	.342	51°	.777	81°	.988
21°	.358	52°	.788	82°	.990
22°	.375	53°	.799	83°	.993
23°	.390	54°	.809	84°	.995
24°	.407	55°	.819	85°	.996
25°	.423	56°	.829	86°	.998
26°	.438	57°	.839	87°	.999
27°	.454	58°	.848	88°	.999
28°	.470	59°	.857	89°	1.000
29°	.485	60°	.866	90°	1.000
30°	.500				

Index